MATHEMATICS RESEARCH DEVELOPMENTS

THE NUMERICAL SOLUTION OF CONTINUOUS TIME OPTIMAL CONTROL PROBLEMS WITH THE CUTTING ANGLE METHOD

MATHEMATICS RESEARCH DEVELOPMENTS

MATHEMATICS RESEARCH DEVELOPMENTS

THE NUMERICAL SOLUTION OF CONTINUOUS TIME OPTIMAL CONTROL PROBLEMS WITH THE CUTTING ANGLE METHOD

SEYEDALIREZA SEYEDI,
IRAJ SADEGH AMIRI,
SARA CHAGHERVAND
AND
VOLKER J SORGER

nova
science publishers
New York

Library of Congress Cataloging-in-Publication Data

ISBN: 978-1-53613-143-7
Library of Congress Control Number: 2018933123

Published by Nova Science Publishers, Inc. † New York

CONTENTS

Preface		**vii**
Chapter 1	Introduction to Cutting Angle Method Inspired by Abstract Convexity for Solving Continuous Time Optimal Control Problems	**1**
Chapter 2	A Wide Literature Review on Building the Cutting Angle Method	**15**
Chapter 3	The Inheritance and Generalizability Properties Extended from Function Definitions into Functionals	**33**
Chapter 4	Study of Some New Type of Functionals Defined Based the Inheritance and Generalizability Properties of Functions	**51**
Chapter 5	Capability of Function Optimization Algorithms for Solving Optimal Control Problems with Respect to the Inheritance and Generalizability Properties	**77**

Chapter 6 A Generalized Version of Cutting Angle Method for Solving Continuous Time Optimal Control Problems **93**

Chapter 7 A Combination of the Cutting Angle Method and a Local Search on Optimal Control Problems **133**

Authors' Contact Information **153**

Index **155**

PREFACE

This book consists of two parts. The first part is on the development of the proposition that "if there exists a type of function, then there exists a functional with the same type" based on the proposition of the inheritance and generalizability properties of a function in a functional. This study presents the abstract convex, increasing positively homogeneous and convex-along-rays functionals via this proposition. The second part concerns the investigation of the use of a global search optimization algorithm called the Cutting Angle Method (CAM) on Optimal Control Problems (OCP). Many algorithms are available for solving OCP, but they are basically local search algorithms. To overcome the problem associated with local searches, most OCP are modeled as Linear Quadratic Regulator (LQR) problems in the hope that the solution found estimates of the true global solution to the original problem. However, in doing so, a lot of information carried by the original problem might be lost in its translation into LQR models. CAM being a global search algorithm is expected to overcome this problem. It can be used alone or in combination with a local search to find the global solution. CAM has been successfully used on functions, however, OCP are functionals. To do this, a model has been introduced based on inheritance and generalizability properties to demonstrate that the optimization algorithms that are used for functions can also be extended for use in functionals. Based on these properties, the study discovered that with the Unit Vectors Combinations Technique (UVCT) proposed in this research, CAM could successfully work on functionals in general and OCP particularly. To help speed up the convergence of CAM, the literature proposed the use of local searches for the determination of the initial solution. In a case study done in the research, CAM was successfully combined with a local search known as the Dynamic Integrated System Optimization and Parameter Estimation (DISOPE) algorithm. Moreover, the initial solution given by the DISOPE algorithm has been verified as a global influence by CAM.

Chapter 1

INTRODUCTION TO CUTTING ANGLE METHOD INSPIRED BY ABSTRACT CONVEXITY FOR SOLVING CONTINUOUS TIME OPTIMAL CONTROL PROBLEMS

ABSTRACT

A survey of optimization techniques for continuous time optimal control problems is given. In particular, we give a brief discussion about a global search so-called "Cutting Angle Method" which comes from abstract convexity. The object of this study is to explain the potential ability of such optimization algorithm to extend for solving a wide range of optimization problems.

Keywords: abstract convexity, cutting angle method, global optimization, continuous time optimal control problems

1.1. INTRODUCTION

Some major tools such as calculus of variations and functional analysis are required to examine and solve Variational problems (VPs)

and in particular Optimal Control Problems (OCP). By and Large, these are the main tools in the study of local optimization of those problems. The gradient and various kinds of generalized derivatives are very useful in the study of local extrema. Nevertheless, local approximation alone cannot help to solve many global optimization problems, thus there is a clear need to develop special global tools for solving this kind of problems.

Although there are no specific algorithms in order to solve any types of optimization problems, numerous optimization algorithms have been generated for solving a few varieties of which. Furthermore, taking a nonlinear aspect of most real-life problems into consideration, a great many optimization algorithms have been developed to solve nonlinear optimization problems classified into two types of global and local optimization algorithms.

Of the two types, the former has the ability to find globally optimal solutions provided that sufficient time is given while the latter, local optimization algorithms, does not. Local optimization algorithms comprise gradient descent, Newton's method, quasi Newton's method, conjugate gradient methods, etc. [1, 2]. From some initial points, these types converge to a local optimum. Such a local optimum is globally optimal providing that the objective function is quasi-convex and viable region is convex, which is rare to occur in practice (method, quasi Newton's method, conjugate gradient methods, etc. [3-8].

Global optimization algorithms embrace mechanisms can avoid local optimization. The focus of this paper is on a type of global optimization called Cutting Angle Method (CAM) which is a generalization of the cutting plane method from convex minimization driven from results on abstract convexity [9]. This method was proposed and later on was developed, in order to minimize the increasing convex-along-rays functions as a special case. Although it can be applied to a very broad class of non-convex global optimization problems in which the functions are involved, possess suitable generalized affine minorants.

1.2. Abstract Convexity Concepts for Defining the Cutting Angle Algorithm

The main area of global and local optimization is convex programming. The basic tool in the study of convex optimization problems is the subgradient. Actually, convex programming plays both a local and global role. First, a subgradient of a convex functional F at a vector $\boldsymbol{x}$ carries out a local approximation of F. Second, the subgradient allows the construction of an affine functional. This affine functional H is called a support functional. Since $F(\mathcal{y}) \geq H(\mathcal{y})$ for all $\mathcal{y}$, the second role is global. In contrast to a local approximation, the functional H shall be called a global affine support [10].

The existence of global affine supports is induced from the renowned separation theorem for convex sets: each element which does not belong to a convex closed set can be separated from this set by a linear functional. This theorem also implies that each lower semicontinuous convex functional F is the upper envelope of the set of all affine functionals. This result is closely correlated to the existence of global affine supports. Because an affine functional is a linear functional plus a constant, then the convexity presents as a combination of linearity and the envelope presentation. Furthermore, in some examples, if an affine global support does not exist, some kind of non-affine global supports can be used. Therefore, the idea arises to study the major concepts of convexity in nonconvex fields as follows,

- abstract convex (AC) functionals as the upper envelope of subsets of all affine functionals.
- increasing positively homogeneous (IPH) and convex-along-rays (CAR) functionals.

Abstract convexity has many applications in the study of functional analysis and optimization. Based on function framework this study shall be focused on increasing positively homogeneous (IPH), abstract convex (AC) and convex-along-rays (CAR) functionals.

1.3. The Cutting Angle Method as a Global Optimization Tool

Optimization problems are made up of three basic ingredients. First is the objective function which should be optimized. Next is a set of unknowns or variables which affect the value of the objective function. Last is a set of constraints that allows the unknowns to take on certain values but exclude others. The optimization problem identifies values of variables that optimize the objective function while satisfying the constraints.

A local minimum is a value of a function which is lower than any nearby values of its argument, or set of values of its arguments. It is a necessary condition for a local minimum of $y = f(x)$ that $\frac{dy}{dx} = 0$; provided this is satisfied, it is a sufficient condition for a local minimum that $\frac{d^2y}{dx^2} > 0$.

In contrast, the global minimum is a value of a function as low as or lower than any other values of its arguments. The sufficient condition for a minimum of a function of a single argument, in terms of a zero first and positive second derivative, shows only that the function takes a local minimum.

Some of the algorithms for nonlinear optimization problems find the global solution, which is the point with the lowest function value among all feasible points. But recognizing the location of the global solution in some applications is difficult. In addition, for convex programming problems (for linear programs particularly) local solution is also a global solution. In contrast, general nonlinear problems, both constrained and unconstrained, may possess local solution that is not the global solution [11].

Optimization algorithms are iterative. They begin with an initial guess of the variable x and generate a sequence of improved estimates (called "iterates") until they approach to a solution. The strategy used to move from one iterate to the next distinguishes one algorithm from another. Most strategies make use of the values of the objective function. Some algorithms accumulate information that is gathered at previous iterations, while others use only local information obtained at the current point. Alternatively, the optimization algorithms check the criterion which is assumed due to the objective function and constraints, at any iteration. When it is satisfied, the optimization algorithms will be stopped and shown the solution as an optimum. Regardless of these specifics, good algorithms should possess the following properties:

- Robustness. They should perform well on a wide variety of problems in their class, for all reasonable values of the starting points.
- Efficiency. They should not require excessive computer time or storage.
- Accuracy. They should be able to identify a solution with precision, without being overly sensitive to errors in the data or to the arithmetic rounding errors that occur when the algorithm is implemented on a computer.

Those goals may conflict together. For example, a rapidly convergent method for a large unconstrained nonlinear problem may require too much computer storage. On the other hand, a robust method may also be the slowest. Tradeoffs between convergence rate and storage requirements, and between robustness and speed, and so on, are central issues in numerical optimization.

In most of the scientific sectors such as physics, aerospace, electrical engineering, mechanical engineering, etc. it is very important to calculate the global optimum value in their equations. Finding significant algorithms with the lowest processing time had priority for engineers and scientists.

The CAM is a deterministic global optimization technique applicable to Lipschitz, AC, IPH and CAR functions. This method, with its roots in the theory of abstract convexity [12], can be viewed from various perspectives: those of Lipschitz approximation, branch-and-bound, tabu search, etc. This research examined the application of CAM as a global search on OCP. Also, as a case study, a combination of CAM with a local search shall be presented.

1.4. The Convex Analysis Tools and the Solution of Optimal Control Problems

The convex analysis is a branch of mathematics that studies convex sets, convex functions (or convex functionals) and convex extremal problems. Some of the main concepts of abstract convexity are to examine AC, IPH and CAR functions which were proposed by Rubinov (2000), Martinez-legaz and Rubinov (2001) and Rubinov and Glover (1996) respectively which is useful for presenting and constructing the AC, IPH and CAR functionals. The notion of convexity is crucial to the solution of many real-world problems. Fortunately, many problems encountered in constrained control and estimation is convex. Convex problems have many important properties for optimization problems.

Moreover, the performance of physical processes depends on a large number of decisions. Optimal control is a study of how to choose the best set of decisions to achieve a particular objective [13]. Physical processes can be divided into two parts. One is the static process and the other one is the dynamic process. The dynamic process can be further divided into two, the linear and the non-linear process. The concern of this study is the non-linear dynamic process.

Optimization of the non-linear dynamic process would model the process as a linear process. Then it is solved iteratively by using an appropriate dynamic optimization algorithm. The solution of the linear process will converge to the solution of the original non-linear problem despite the inaccuracies of the mathematical models used [14, 15].

Whereas, many algorithms are available for solving OCP but they are basically local search algorithms. To overcome the problem associated with local searches, most OCP is modeled as Linear Quadratic Regulator (LQR) problems in the hope that the solution found estimates the true global solution to the original problem. However, in doing so, a lot of information carried by the original problem might be lost in its translation into LQR models.

This study shall propose a dynamic global search algorithm called the Cutting Angle Method (CAM) which was introduced by Bagirov and Rubinov (2000) as a generalization of the cutting plane method from convex minimization, to solve OCP. CAM has been successfully used for functions, however, OCP is functionals.

The CAM can be applied in a very broad class of non-convex global optimization problems. Bagirov and Rubinov (2000) claimed CAM as able to help local search algorithms to converge to global optima. This method is used in combination with a local search algorithm to arrive at a truly global solution [16]. Basically, the approach begins with a solution of a local search algorithm. The cutting angle method is then applied and repeated until one arrives at a solution that is lower than the current solution of the local search algorithm. Using the lower solution as the new initial solution, the local search algorithm is again run on the problem. Once a new solution is gotten from this local search, the cutting angle method is once again applied. The process would stop only when the cutting angle method fails to find a lower valued solution after a predetermined number of iterations. The most recent solution of the local search algorithm would be considered as the global optimum.

As a case study, this research proposes to use a local search algorithm to go together with the CAM. This study has chosen the Dynamic Integrated Systems Optimization and Parameter Estimation (DISOPE) algorithm as the local search. This algorithm is an algorithm specially written for solving OCP. It uses LQR models as input and the basic gradient as search direction. The CAM used the special transformed objective function for each intermediate use of the method. In their papers, Bagirov and Rubinov (2003) used real-valued functions for the objective functions in their solved problems. The main concern of this research is to introduce a technique to use CAM for solving OCP directly and with a combination with a local search.

1.5. The Scope of the Study of Solving Optimal Control Problems with Cutting Angle Method

The main aim of this study is considering the ability of CAM to locate the global solution for OCP with a chosen local search. The research will basically be answering the following questions:

- How to propose main tools from abstract convexity?
- How to create a transformed function for a functional?
- Is CAM suitable for optimal control problems (OCP)?
- How to generate the initial solution for CAM?
- Is CAM compatible with DISOPE as a local search?

That is, the principle objectives of this study are:

- To explore and gather the basic tools from the abstract convex analysis, functional analysis, and optimal control theory.
- To create transformed functions suitable for functionals.

- To study the cutting plane method as basis for the cutting angle method
- To use CAM directly on OCP
- To study DISOPE algorithm as a local search.
- To generate the initial solution for CAM.
- To use a combination of CAM and DISOPE algorithm on OCP.

This research demonstrates AC, IPH and CAR functionals based on the introduction of Inheritance and Generalizability function properties on functional. Moreover, this study covers the ability of CAM to solve OCP based on the performance of capability of function optimization algorithms to solve OCP directly via unit vectors combinations technique (UVCT). Also, DISOPE algorithm will be used as a local search to generate the initial solution for CAM. This study will be a significant endeavor in applying the CAM to optimize functionals in general and OCP in particular. In this manner, this study will also be a major effort in introducing new types of functionals via the Inheritance and Generalizability properties and to utilize the optimization algorithms which are used for calculus problems on the calculus of variations problems (and OCP particularly). Moreover, this research will provide recommendations on how to approximate the optimal control problem with a combination of approximation and discretization methods.

This study will be beneficial to the students and researchers in functional analysis and optimization when they work with functionals with functions framework. It will also serve as a future reference for researchers on the subject of functional analysis, calculus of variations, optimal control, and optimization algorithms.

This study organizes as follow:

Chapter 1 gives a brief introduction and scope of the work "Application of the Cutting Angle Method in Optimal Control Problems". Chapter 2 gives literature reviews on published relevant works. This chapter begins with the introduction of abstract convexity and optimal control. Then the history of original CAM is presented. Chapter 3 gives a model based on the Inheritance and Generalizability properties in both functions and functionals. Based on the properties, the following assertion is investigated:

If there is any type of function then there exists its same-name functional i.e., if there is the convex function then there exists the convex functional.

Chapter 4 reports an introduction of some new types of functionals i.e., AC functionals. In Chapter 5, the existence of the Inheritance and Generalizability properties in optimization algorithms is investigated. These properties verified that optimization algorithms which can be used on functions can also be used on functionals. Later, in Chapter 6, a direct cutting angle method for OCP is proposed based on a new technique and discussed. In Chapter 7 a combination of CAM and a local search is demonstrated as a case study. In addition, the last chapter gives some suggestions for future work.

1.6. Research Methodology of This Work

This research consists of two basic phases. One phase occurs in abstract analysis field and another one occurs in optimization area. Each phase has two subsections. One is theoretical part and another is a numerical part. The theoretical part prepares appropriate tools for validating numerical part. Furthermore, any validation processes in optimization phase were simulated from abstract analysis phase. Follow is the flowchart for each step in the methodology of the research:

Moreover, in the above flowchart, the darker rectangular frames are some of the results of the research, whereas the lighter rectangular frames are gathered from the literature. The two phases of this accomplished work are:

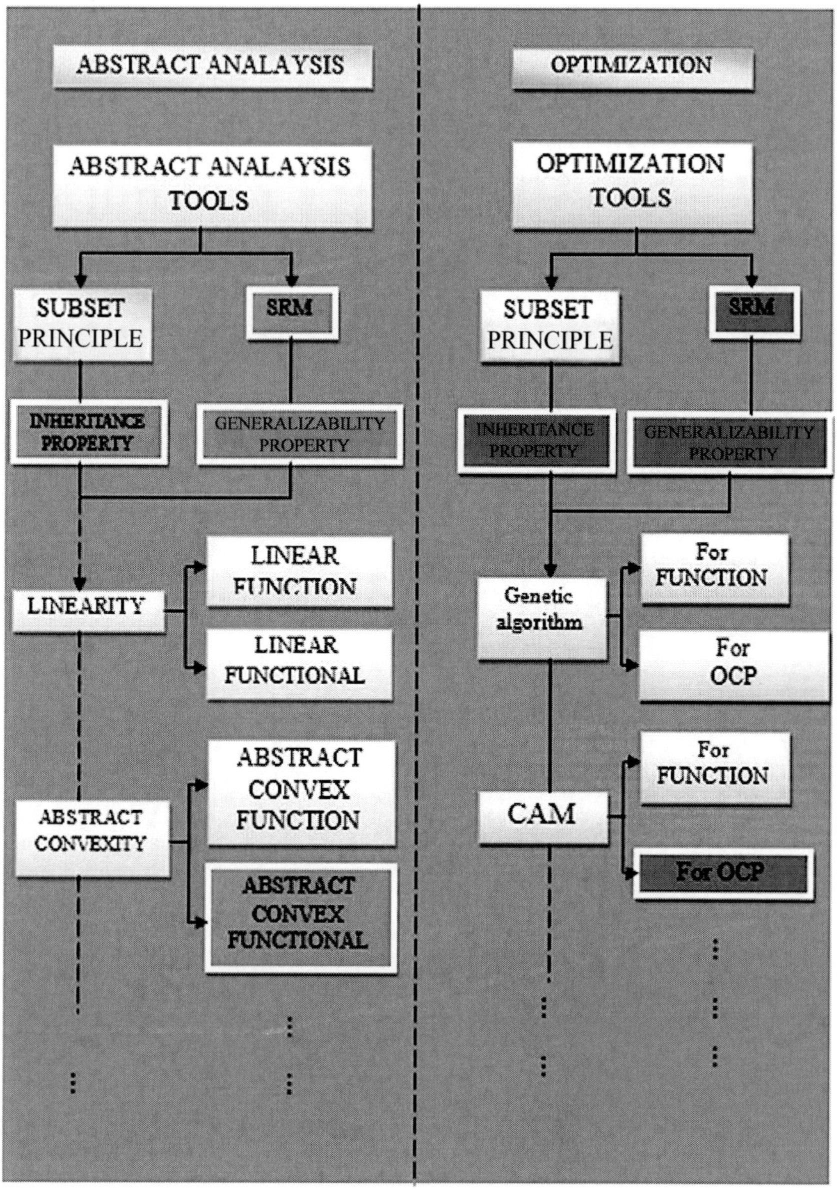

Figure 1. Research Methodology flowchart.

1.6.1. Phase 1: Abstract Analysis

The first theoretical part supports that if there is any type of function on Euclidean space then there exists its same-name functional i.e., if there is the convex function then there exists the convex functional based on a model to be introduced. Based on this assertion, since AC, IPH and CAR functions exist thus AC, IPH and CAR exist.

1.6.2. Phase 2: Optimization

The second theoretical part support that any optimization algorithms which can be used on functions also can be used on functionals. It implies that since CAM was used for functions successfully thus it can also be used for functionals. This research demonstrated an application of CAM on OCP successfully via unit vectors combinations technique (UVCT). Moreover, this study investigates a successful combination of CAM and a local search (DISOPE algorithm) as a case study.

REFERENCES

[1] Baldi, P. (1995). “Gradient descent learning algorithm overview: A general dynamical systems perspective”, *Neural Networks, IEEE Transactions on*, *6*(1), 182-195.

[2] Shang, Y. (1997). *“Global search methods for solving nonlinear optimization problems,”* University of Illinois at Urbana-Champaign.

[3] Beale, E. M. L. (1988). *Introduction to optimization*: Wiley-Interscience.

[4] Bundy, B. (1984). *“Basic optimization methods”*, Edward Arnold.

[5] Cesari, L. (1983). *Optimization—theory and applications*: Springer.
[6] Koo, D. (1977). *Elements of optimization*: Springer.
[7] Rao, S. "*Optimization: theory and applications. 1984*," ed: Wiley, New York.
[8] Stoer, J. & Witzgall, C. (1970). *Convexity and optimization in finite dimensions*: Springer.
[9] Rubinov, A. M. (2000). *Abstract convexity and global optimization*, vol. 44, Springer.
[10] Kurdila, A. J. & Zabarankin, M. (2005). "*Convex Functional Analysis, Systems and Control: Foundations and Applications*," ed: Birkhäuser, Basel.
[11] Nocedal, J. & Wright, S. J. (2006). "*Numerical Optimization 2nd*".
[12] Rubinov, A. & Glover, B. (1999). "Increasing convex-along-rays functions with applications to global optimization", *Journal of Optimization Theory and Applications*, *102*(3), 615-642.
[13] Singh, M. G. & Titli, A. (1987). *Systems: decomposition, optimisation, and control*: Pergamon.
[14] Becerra, V. M. (1994). *"Development and applications of novel optimal control algorithms,"* City University.
[15] Roberts, P. (1993). "An algorithm for optimal control of nonlinear systems with model-reality differences", in *Proceedings of 12th IFAC World Congress on Automatic Control*, 407-412.
[16] Bagirov, A. & Rubinov, A. (2000). "Global minimization of increasing positively homogeneous functions over the unit simplex", *Annals of Operations Research*, 98(1-4), 171-187.

Chapter 2

A Wide Literature Review on Building the Cutting Angle Method

Abstract

A broad study of Abstract Convexity is initiated. One of the main applications of this concept is global optimization. An essential result of such applications is a global search so-called Cutting Angle Method (CAM). We give a brief discussion about the structure and combination feature of this algorithm. In addition, at the end of this study a brief description of another powerful search called "Dynamic Integrated Systems Optimization and Parameter Estimation (DISOPE) Algorithm" is given.

Keywords: abstract convexity, cutting angle method, dynamic integrated systems optimization and parameter estimation algorithm

2.1. Introduction

Literature reviews of the subject matters of this research are given in this chapter. Some of the discussions in Chapter 1 are given elaboration here.

First, AC, IPH [1] and CAR functions shall be reviewed as the basic concepts of abstract convex which would be useful for proposing their same-name functionals i.e., the same-name of IPH function on functional is IPH functional.

Then, the history of the conceptions of CAM (as a global dynamic search) and DISOPE (as a local dynamic search) shall be discussed.

2.2. Abstract Convexity Concepts Defined on the Cone of Euclidean Space

Suppose $\mathcal{A}, \mathcal{B}, \mathcal{C}$ and $\mathcal{D}$ are sets. The ordered pair $(\mathcal{A}, \mathcal{B})$ is defined by the set $\{\{\mathcal{A}\}, \{\mathcal{A}, \mathcal{B}\}\}$. Furthermore, the ordered triple $(\mathcal{A}, \mathcal{B}, \mathcal{C})$ is $((\mathcal{A}, \mathcal{B}), \mathcal{C})$ and the ordered quadruple $(\mathcal{A}, \mathcal{B}, \mathcal{C}, \mathcal{D})$ is $((\mathcal{A}, \mathcal{B}, \mathcal{C}), \mathcal{D})$. As well, a set is called a relation if and only if all its members are ordered pairs. Also, a set f is called a function if and only if f is a relation and for each $x \in domain(f)$ there exists a unique set $\mathcal{Y}$ such that $\mathcal{X}, \mathcal{Y} \in f(x, y)$, (Searcoid, 1948). Traditionally a functional is a function that takes functions as its argument or input and returns a real number. An important application of functional is in calculus of variations where one searches for a function which minimizes a certain functional (Todd, 1999 and Lang, 1993).

Let $\tilde{Q}$ be either $\mathbb{R}^n_+$ or $\mathbb{R}^n_{++}$. The main results of convex analysis state the following definitions which would be useful for presentation of AC, IPH and CAR functionals in Chapter 5 as their same-name extensions on functional.

Definition 2.1: Let $\mathcal{V} \subset \overline{\mathbb{R}}$ and let $\mathcal{H}$ be a nonempty set of functions $h: \mathcal{X} \to \mathcal{V}$. A function $h: \mathcal{X} \to \overline{\mathcal{V}}$ is called abstract convex with respect to H (or H-convex) if there exists a set $\mathcal{U} \subset \mathcal{H}$ such that f is the upper envelop of this set:

$$f(x) = \sup\{h(x): h \in \mathcal{U}\} \text{ for all } x \in \mathcal{X}$$

Definition 2.2: Let $\tilde{Q}$ be either $\mathbb{R}^n_+$ or $\mathbb{R}^n_{++}$. A function $F: \tilde{Q} \to \mathbb{R}_{+\infty}$ is an IPH function if the following conditions are satisfied:

1. $x \geq y$ implies that $F(x) \geq F(y)$:
2. $(bx) = bF(x)$ for all $x \in \tilde{Q}$ and $b > 0$ and $b \in \mathbb{R}_{+}$,.

Definition 2.3: A function $F: Q \to \mathbb{R}_{+\infty}$ is an ICAR function if the following conditions hold:

- F is increasing: $x \geq y$ implies $F(x) \geq F(y)$;
- for each $x \in Q$, the functional F_x defined by

$$F_x(t) = F(tx) \qquad t \in (0,1),$$

is convex.

Abstract convexity has found many applications for optimization that shall be discussed in the next section.

2.3. The Cutting Angle Method as a Main Global Search Defined on Abstract Convexity

Optimization may be defined as the science of determining the "best" solutions to certain mathematically defined problems, which are often models of physical reality. The field involves the study of optimality criteria for problems, algorithms and solution methods, the study of structures of such methods and computer simulations with the methods both using benchmark problems and/or real life problems [2].

Optimization is the fact of obtaining the best result under given circumstance. In human kinds' daily lives, a lot of technological and managerial decisions are made. The ultimate goal of all such decisions is to either optimize the effort required or the desired benefit. Since the effort required or the benefit desired in any practical situation can be expressed as a function of certain decision variables, optimization can be defined as the process of finding the conditions that give the optimum value of a function.

In general, optimization problems consisted of three basic components. They are a set of unknowns or variables, a set of constraints that specify the feasible values of the variables, and an objective function to be optimized. The optimization problem involves finding the values of the variables that optimize the objective function while satisfying the constraints.

There are many types of optimization problems. Basically, the can be categorized as continuous or discrete, linear or nonlinear and constrained or unconstrained depending on the forms of their components. A problem is continuous if the variables take on continuous real values; otherwise, it is a discrete problem. The variables of a discrete problem usually take on the integer values.

A problem is linear if both the objective function and the constraint are linear. When either of them or both of them are nonlinear, the problem is nonlinear. Most of real life problems are nonlinear. Optimization problems may be constrained or unconstrained depending on the existence of conditions imposed on the variables as the optimization process take place. An unconstrained optimization problem is uni-modal if its objective function is convex. A uni-modal optimization problem has one local optimum that is also global optimum. Uni-modal problems are relatively easy to solve. Methods such as gradient descent, Newton's, quasi Newton's and conjugate gradients are readily available for finding their solutions.

Generally, nonlinear optimization problems are difficult problems. The problems are multi-modal. Multi-modal problems are problems where the objective functions have many local optima. These local optima might not be the global optima. Furthermore, the nonlinear constraints themselves are challenging because they may represent feasible regions that are difficult to find.

There is no single algorithm available for solving all optimization problems efficiently. Hence, a number of optimization algorithms have been developed for solving these different types of optimization problems. The existence of optimization algorithms can be traced to the days of Newton, Lagrange and Cauchy. The development of differential calculus methods for optimization was made possible through the contributions of Newton and Leibnitz. Bernoulli, Euler, Lagrange and Weirstrass laid the foundations for the calculus of variations [3]. The algorithm of optimization for constrained problems, which involves the addition of unknown multipliers, became known by the name of its inventor, Lagrange. Cauchy made the first application of the steepest descent algorithm to solve unconstrained optimization problems.

In spite of these early contributions, very little progress was made until the middle of the twentieth century, when high-speed computers made the implementation of the optimization procedures possible and stimulated further research on new algorithms. This advancement resulted in the emergence of massive literature on optimization techniques and several well-defined new areas in optimization theory.

As mentioned earlier, most real life problems are nonlinear problems. Hence, many optimization algorithms have been developed to solve nonlinear optimization problems. Nonlinear optimization algorithms are classified into local optimization and global optimization algorithms. Global optimization algorithms can find global optimal solutions given long enough time, while local optimization algorithms do not.

Local optimization algorithms include gradient descent, Newton's method, quasi Newton's method and conjugate gradient methods [4]. They converge to a local optimum from some initial points. Such a local optimum is globally optimal only when the objective function is quasi-convex and feasible region is convex, which rarely happens in practice [5-8]. To overcome local optima and search for global optima, global optimization algorithms have been developed. These optimization algorithms have mechanisms that can escape local optima.

This research focuses on dynamic optimization, better known as optimal control. It concerns the study of optimization problems in which the governing constraints are dynamical in nature. The constraints are normally represented by systems of equations, which may be continuous or discrete. An important use of optimal control is for finding optimal trajectories for nonlinear dynamical systems. The problem addressed in this study would be the problems of finding the optimal control of nonlinear dynamical systems.

CAM is based on results in abstract convexity. The CAM arises, as do the Piyavskii and Mladineo methods [9, 10]. It was proposed by Andramonov et al. (1999) and extended by Rubinov (2000) for minimizing the increasing convex-along-rays functions as a special case, CAM is a generalization of the cutting plane method (CPM) from convex minimization. It can be applied in a very broad class of non-convex global optimization problems in which the functions involved, possess suitable generalized affine minorants. The following procedure of definition of CAM was proposed by Batten and Beliakov (2002).

Consider the following global optimization problem:

$$f(x) \to min$$
$$\text{subject to } x \in \mathcal{S}^*,$$

where $\mathcal{S}^* = \{x \in \mathbb{R}^n_+ : \sum_i^r x_i = 1\}\}$ and r is the number of coordinates of the vector x.

This problem can be reformulated as the global optimization problem of the so-called IPH function over the unit simplex. The class of IPH functions f defined on $\mathbb{R}^n_+$ is

$$\{f : \forall x, y \in \mathbb{R}^n_+ \; x \geq y \text{ implies } f(x) \geq f(y); \; \forall x \in \mathbb{R}^n_+, b \in \mathbb{R}^n_+, b > 0:$$

$$f(bx) = bf(x)\},$$

where $\mathbb{R}_+^n$ denotes the set of all real vectors with non-negative components.

Let $g: S^* \to \mathbb{R}$ be a positive Lipschitz function defined on the unit simplex. Then it can be extended to a finite IPH function $f(x)$ on the cone $\mathbb{R}_+^n$, which would coincide with $g(x) + c$ on S^*. i.e., $f(x) = g(x) + c$ is an IPH function on S^*. Then, for each $x \in \mathbb{R}_+^n$ define the support vector

$$l = \left(\frac{f(x)}{x}\right) = \left(\frac{f(x)}{x_1}, \frac{f(x)}{x_2}, \dots, \frac{f(x)}{x_n}\right).$$

The n vectors $e^m = (0, \dots, 0, 1, 0, \dots, 0)$, with 1 in the mth position shall be used, and the corresponding support vectors $l^m = \left(\frac{f(e^m)}{e^m}\right)$, $m = 1, \dots, n$ shall be called basis vectors. A set of $k \geq n$ support vectors (and hence k known values of the function $f(x)$ at k distinct points), $\mathcal{K} = \{l^k\}_{j=1}^{k}$. Let also the first n support vectors be the basis vectors. This choice of support vectors guarantees that the algorithm will locate all local (and hence global) minimizers of the auxiliary function $h_k(x) = \max_{j \geq k} \min_{i=1,\dots,n} l_i^j x_i$ in the interior of the unit simplex. It always underestimates the value of $f(x)$: $h_k(x) \leq f(x)$. Hence, $b_k = \min_{x \in S^*} h_k(x) \leq \min_{x \in S^*} f(x)$. On the other hand, the sequence of its minima, $\{b_k\}_{k=n}^{\infty}$ is increasing [11], and convergence to the global minmum of $f(x)$ and the CAM was formulated by Rubinov (2000) as follows.

Algorithm 1

Step 0. Initialization

(a) Take points e^m, $m = 1, \dots, n$, and construct basic vectors $l^m = \left(\frac{f(e^m)}{e^m}\right)$, $m = 1, \dots, n$.

(b) Define the function $h_n(x) = \max_{j \le n} \min_{i=1,\dots,n} l_i^j x_i = \max_{j \le n} l_j^j x_j$.

(c) Set $k = n$.

Step 1. Find $x^* = \arg[\min_{x \in S^*} h_k(x)]$.

Step 2. Set $k = k + 1$ and $x^k = x^*$.

Step 3. Compute $l^j = \left(\frac{f(x^k)}{x^k}\right)$. Define the function

$$h_k(x) = \max_{j \le k} \min_{i=1,\dots,n} l_i^j x_i = \max\{h_{k-1}(x), \min_{i=1,\dots,n} l_i^k x_i\}.$$

Go to Step 2.

A more general version of this algorithm is known as the ϕ-bundle method, and its convergence under very mild assumptions was proven by Pallachke and Rolewicz (1997). The crucial and most time-consuming step of the Algorithm 1 is Step 1, minimization of the auxiliary function. This problem is essentially of combinatorial nature. Some properties of the auxiliary function are studied by Bagirov and Rubinov (2000). Among them the following is noted.

Theorem 2.1: *Let $x > 0$ be a local minimizer of $h_k(x)$ over the relative interior of S^*, $riS^* = \{x \in S^*, x > 0\}$. Then there exists a subset $\mathcal{L} = \{l^{j_1}, l^{j_2}, \dots, l^{j_n}\}$ of the set $\mathcal{K}$, such that*

1. $x = \{d/l_1^{j_1}, d/l_2^{j_2}, \dots, d/l_n^{j_n}\}$ with $d = (\sum_i 1/l^{j_i})^{-1}$.
2. $\max_{j \le k} \min_{i=1,\dots,n} l_i^j / l_i^{j_i} = 1$.
3. *Either* $\forall i: j_i = i$, *or* $\exists m: k_m > n, l_i^{j_m} > l_i^{j_i}, \forall i \ne m$.

The value of the auxiliary function at x is $h_k(x) = d$.

Remark 2.1: Bagirov and Rubinov (2003) also proved that $\forall i, j, 1 \leq n, n+1 \leq j \leq k$: $l_i^i \leq l_i^j$.

In order to find the global minimum of the auxiliary function at Step 2 of the algorithm, all its local minima are needed to examined, and hence all combinations of the support vectors that satisfy the conditions of the Theorem 2.1. This process can be significantly accelerated by noticing, that

$$h_k(x) = \max\{h_{k-1}(x), \min_{i=1,\dots,n} l_i^k x_i\}.$$

Then, if we have already computed all the local minima of the auxiliary function $h_{k-1}(x)$ at the previous iteration, those minima that have been added by aggregation of the last support vector l^k. This means that examination of only those combinations of supports vectors that include vector l^k is needed (i.e., one of $l^{j_i} = l^k$). The following version of CAM that works based on the above theorem, by examining all possible combinations of n support vectors was improved by Batten and Beliakov (2002).

2.4. Combinatorial Formulation of the Cutting Angle Method on Abstract Convexity

This section translates the crucial step of the CAM, minimization of the auxiliary function $h_k(x)$, into an abstract combinatorial problem of selection of groups of n support vectors which was proposed by Batten and Beliakov (2002). Consider a set of k support vectors $\mathcal{K} = \{l^j\}_{j=1}^k$, $l^j \in \mathbb{R}_+^n$. Let $\mathcal{I}$ denote $\{1,2,\dots,n\}$. From Theorem 2.1, the local minima of the auxiliary function $h_k(x)$ are combinations of n support vectors $\mathcal{L} = \{l^{j_1}, l^{j_2}, \dots, l^{j_n}\}$ that satisfy the following conditions:

(I) $\forall i, s \in \mathcal{J}, i \neq s\colon l_i^{j_i} < l_i^{j_s}$

(II) $\forall v \in \mathcal{K}\backslash\mathcal{L}, \exists i \in \mathcal{J}\colon l_i^{j_i} \geq v_i$.

The subset $\mathcal{L}$, which satisfied conditions (I) and (II) above, is called a valid combination of support vectors.

To illustrate these conditions, interpret $\mathcal{L}$ as a $n \times n$ matrix, whose rows are $l^{j_1}, l^{j_2}, \dots, l^{j_n}$:

$$\mathcal{L} = \begin{pmatrix} l_1^{j_1} & l_2^{j_1} & \cdots & l_n^{j_1} \\ l_1^{j_2} & l_2^{j_2} & \cdots & \vdots \\ \vdots & \vdots & \vdots & \vdots \\ l_1^{j_n} & \cdots & \cdots & l_n^{j_n} \end{pmatrix}.$$

The element of this matrix will be denoted by $k_{si} = l_j^{k_s}$. Condition (I) implies that every element on the diagonal must be the smallest in its column, and condition (II) implies that for every vector v of $\mathcal{K}$ that is taken, not already in $\mathcal{L}$, the diagonal of $\mathcal{L}$ is not dominated by v: $\neg(diag(\mathcal{L}) = (l_1^{j_1}, \dots, l_n^{j_n}) < v)$.

An example, consider the set $\mathcal{K} = \{(1,\infty,\infty), (\infty,2,\infty), (\infty,\infty,2), (2,3,4), (3,4,3)\}$. List the combinations satisfying (I):

$$\mathcal{L}_1 = \{l^1, l^2, l^3\},\ \mathcal{L}_2 = \{l^4, l^2, l^3\},\ \mathcal{L}_3 = \{l^1, l^4, l^3\},$$

$$\mathcal{L}_4 = \{l^1, l^2, l^4\},\ \mathcal{L}_5 = \{l^5, l^2, l^3\},\ \mathcal{L}_6 = \{l^1, l^5, l^3\},$$

$$\mathcal{L}_7 = \{l^1, l^2, l^5\},\ \mathcal{L}_8 = \{l^4, l^2, l^5\},\ \mathcal{L}_9 = \{l^1, l^4, l^5\}.$$

Choose combinations satisfying (I) and (II):

$$\mathcal{L}_4 = \{l^1, l^2, l^4\} = \begin{Bmatrix} 1 & \infty & \infty \\ \infty & 2 & \infty \\ 2 & \infty & 4 \end{Bmatrix}, \mathcal{L}_5 = \{l^5, l^2, l^3\} = \begin{Bmatrix} 3 & 4 & 3 \\ \infty & 2 & \infty \\ \infty & \infty & 2 \end{Bmatrix},$$

$$\mathcal{L}_6 = \{l^1, l^5, l^3\} = \begin{Bmatrix} 1 & \infty & \infty \\ 3 & 4 & 3 \\ \infty & \infty & 2 \end{Bmatrix}, \mathcal{L}_8 = \{l^4, l^2, l^5\} = \begin{Bmatrix} 2 & 3 & 4 \\ \infty & 2 & \infty \\ 3 & 4 & 3 \end{Bmatrix},$$

$$\mathcal{L}_9 = \{l^1, l^4, l^5\} = \begin{Bmatrix} 1 & \infty & \infty \\ 2 & 3 & 4 \\ 3 & 4 & 3 \end{Bmatrix}.$$

Let $\mathcal{W}^{k}$ denote the set of all valid combinations $\mathcal{L}$ of k support vectors satisfying conditions (I) and (II):

$$\mathcal{W}^{k} = \{\mathcal{L} = \{l^{j_1}, l^{j_2}, \dots, l^{j_n}\}, l^{j_i} \in \mathcal{K}: (1)\forall i, s \in \mathcal{I}, i \neq s: \mathcal{L}_{ii} < \mathcal{L}_{si} \text{ and } (2)\ \forall v \in \mathcal{K} \backslash \mathcal{L}\ \exists i \in \mathcal{I}: \mathcal{L}_{ii} \geq v_i\}$$

The problem of finding local minima of $h_k(x)$ is translated into the problem of listing the elements of $\mathcal{W}^{k}$. A simplistic approach is then:

Step 1. Construct all combinations $\mathcal{L}$ satisfying (I)
Step 2. Check the obtained combinations against (II).

Rubinov (2000) and Bagirov and Rubinov (2000) then improve on this by taking into account the fact that all elements of $\mathcal{W}^{k}$ that do not involve l^{j} (i.e., minima of $h_{k-1}(x)$) do not need to be recomputed, hence the algorithm takes the form:

Algorithm 2

Step 0. Initialization

(a) Take points e^m, $m = 1, \ldots, n$, and construct basic vectors $l^m = \left(\frac{f(e^m)}{e^m}\right)$, $m = 1, \ldots, n$.

(b) Define the function $h_n(x) = \max_{j \le n} \min_{i=1,\ldots,n} l_i^j x_i = \max_{j \le n} l_j^j x_j$.

(c) Set $k = n$. Set $\mathcal{W}^k = \{\{l^1, l^2, \ldots, l^n\}\}$.

(d) Calculate $d = (\sum_{i=1,\ldots n} 1/l_i^i)^{-1}$.

Step 1.

(a) Retrieve all valid combinations $\mathcal{L}$ (i.e., $\mathcal{W}^k$).

(b) Select $\mathcal{L} \in \mathcal{W}^k$.

Step 2.

(a) $k = k + 1$.

(b) Take $x^* = d/\text{diag}(\mathcal{L})$ and evaluate $f(x^*)$.

(c) Compute $l^k = (f(x^*)/x^*)$. Define the function

$$h_k(x) = \max_{j \le k} \min_{i=1,\ldots,n} l_i^j x_i = \max\{h_{k-1}(x), \min_{i=1,\ldots,n} l_i^k x_i\}$$

Step 3.

(a) Check $\mathcal{W}^{k-1}$ against (II) and remove those that fail (II).

(b) Move the remaining combinations into $\mathcal{W}^k$.

Step 4.

(a) Construct all combinations $\mathcal{L}$ that involve l^k and satisfy (I).

(b) Calculate $d = (\sum_{i=1,\ldots n} 1/l_i^{k_i})^{-1}$ for each such combination.

(c) Add these combinations to $\mathcal{W}^k$.

(d) Go to Step 1.

It is clear that at Step 4, the number of possible combinations $\mathcal{L}$ that formally need to be constructed and checked at each iteration p is $n\binom{p-1}{n-1}$, where $\binom{a}{c}$ denote binomial coefficients. Since $o(n)$ operations are needed to test condition (I), the complexity of the algorithm is$o(\binom{p-1}{n-1}n^2)$. Of course, in practice, fewer operations are needed: if, when forming $\mathcal{L}$, condition (I) fails at half-way, there is no need to complete the construction of this $\mathcal{L}$ in order to discard it. Still, the complexity of Step 4 is huge. The complexity of Step 3 of the algorithm is $o(|W^{k-1}|kn)$, where $|W^{k-1}|$ is the cardinality of W^{k-1}, i.e., the number of local minima of $h_{k-1}(x)$.

By and large, the CAM was used successfully for AC, IPH and CAR functions by Andramonov, et al., Bagirov and Robinov (2000), Batten and Beliakov (2002) and this research shall investigate an application of the method for solving OCP for the first time.

2.5. A Local Search in Optimal Control Theory So-Called Dynamic Integrated Systems Optimization and Parameter Estimation (DISOPE) Algorithm

Optimal control theory plays an increasingly important role in the design of modern systems. Its objectives are the optimization of the return from, or the optimization of the return from, or the optimization of the cost of, the operation of physical, social and economic processes. The objective of the optimal control theory is to determine the control signals that will cause a process to satisfy the physical constraints and at the same time optimize some performance criterion.

Optimal control has grown rapidly over the years. Rigorous mathematical analysis has put the subject on a sound theoretical basis. Constructive methods have been developed to allow solutions to well-posed problems to be found. The mathematical modeling of optimal control problems arising in a wide variety of contexts has been achieved. It is recognized as an important tool for the solution of problems that occur naturally in such diverse fields as medicine, dynamics, ecology, economics, oil recovery and electric power production [12].

The essential properties of an optimal control problem are the system that evolves in time according to certain laws. These laws are embodied in equations that contain elements, which can be adjusted from outside the system, known as the controls. With suitable choice of these controls, it may be possible to force the system into a desired target state. If this can be done at all, it can usually be done in many different ways. The choice among the successful controls of the optimal ones is governed by the necessity to make some quantity, known as the cost, as small as possible [13]. Optimal control problems can further be categorized as continuous or discrete problems. The concern of this study is only with the continuous optimal control problems.

Discretization of the objectives and constraints is the first step of solving OCP while the function optimization algorithms are used directly. Some researchers developed some discretization methods to find the best approximation for original problem on optimal control i.e., Kumiko (1980) discretized the original problem by Chebyshev series, Razzaghi (1990) by Fourier series, Fraser-andrews (1996) by Shooting method, Jaddu and Shimemura (1999) and Jaddu (2002) by Chebyshev polynomials, Levin et al. (2001) by a sum of squares and, Fard and Borzabadi (2007) by quasi-assignment problem. The appropriate discretization of original OCP will be helpful to achieve the optimal solution. In this research, the Simpson Rule has been used in the discretization OCP.

The DISOPE algorithm in this research is an extension of an earlier algorithm called Integrated System Optimization and Parameter Estimation (ISOPE). It was originally developed by P.D. Roberts (1979), and Roberts and Williams (1981) for online steady-state optimization of industrial processes implemented through adjustment of regulatory controller set points. It has been proved to be successful in solving many example problems [14-17]. Later, Brdys and Roberts (1987) derived sufficient conditions for convergence of this algorithm.

An essential feature of ISOPE is that the iterations converged to the correct real optimum. ISOPE was intended for the steady-state optimizing control. Naturally, it was later extended to solve the dynamical optimal control problems. Roberts (1994) extended ISOPE to dynamical problems and it has been termed very much the same. As it was originally developed and published, DISOPE addressed continuous-time, unconstrained, centralized and time invariant optimal control problems [18]. Becerra (1996) advanced and improved the existing knowledge on the technique so as to make it attractive for its implementation in the process industry.

DISOPE was initially developed for continuous nonlinear optimal control problems [19] and then extended to discrete systems [18] and to optimal tracking control problems [20]. The technique has also been extended to cope with un-matched terminal constraints. A range of application of DISOPE techniques has also been developed by Becerra (1994),

The method is iterative in nature. Repeated solutions of optimization and estimation of parameters within the model is used for calculating the optimum. An important property of the model is that the iterations converge to the real optimum. An implementable algorithm based on linear quadratic regulator (LQR) has been designed and implemented in MATLAB by Roberts (1993). The algorithm integrates the information from the real problem and its simplified model by introducing parameters such that the solution of the model provides the control as a function of the current parameter estimates.

The CAM was combined successfully with the Gradient Descent Method (GDM) for functions by Bagirov and Robinov (2003). In this research, the combination of CAM and DISOPE algorithm for OCP shall be introduced for the first time.

REFERENCES

[1] Martinez-Legaz, J. & Rubinov, A. (2001). "Increasing positively homogeneous functions defined on Rn", *Acta Math. Vietnam*, *26*(3), 313-331.

[2] Fletcher, R. (2013). *Practical methods of optimization*: John Wiley & Sons.

[3] Rao, S. "*Optimization: theory and applications. 1984*," ed: Wiley, New York.

[4] Baldi, P. (1995). "Gradient descent learning algorithm overview: A general dynamical systems perspective", *Neural Networks, IEEE Transactions on*, *6*(1), 182-195.

[5] Beale, E. M. L. (1988). *Introduction to optimization*: Wiley-Interscience.

[6] Bundy, B. (1984). *"Basic optimization methods"*, Edward Arnold.

[7] Cesari, L. (1983). *Optimization—theory and applications*: Springer.

[8] Stoer, J. & Witzgall, C. (1970). *Convexity and optimization in finite dimensions*: Springer.

[9] Hansen, P. & Jaumard, B. (1995). *Lipschitz optimization*: Springer.

[10] Mladineo, R. H. (1986). "An algorithm for finding the global maximum of a multimodal, multivariate function", *Mathematical Programming*, 34(2), 188-200.

[11] Bagirov, A. M. & Rubinov, A. M. (2003). "Cutting angle method and a local search", *Journal of Global Optimization*, 27(2-3), 193-213.

[12] Siouris, G. M. (1996). *An engineering approach to optimal control and estimation theory*: John Wiley & Sons, Inc.

[13] Hocking, L. (2001). "*Optimal control. An introduction to the theory with applications*." Clarendon, ed: Oxford.

[14] Brdyś, M. & Roberts, P. (1987). "Convergence and optimality of modified two-step algorithm for integrated system optimization

and parameter estimation", *International journal of systems science*, 18(7), 1305-1322.

[15] Ellis, J. & Roberts, P. (1981). "Simple models for integrated optimization and parameter estimation", *International Journal of Systems Science*, 12(4), 455-472.

[16] Roberts, P. & Williams, T. (1981). "On an algorithm for combined system optimisation and parameter estimation", *Automatica*, 17(1), 199-209.

[17] Stevenson, I., Brdys, M. & Roberts, P. (1985). "Integrated system optimization and parameter estimation for travelling load furnace control", in *Proceedings of 7th IFAC Symposium on Identification and Parameter Estimation*, 1641-1646.

[18] Becerra, V. & Roberts, P. (1996). "Dynamic integrated system optimization and parameter estimation for discrete time optimal control of nonlinear systems", *International Journal of Control*, 63(2), 257-281.

[19] Roberts, P. (1993). "An algorithm for optimal control of nonlinear systems with model-reality differences", in *Proceedings of 12th IFAC World Congress on Automatic Control*, 407-412.

[20] Becerra, V. & Roberts, P. (1998). "Application of a novel optimal control algorithm to a benchmark fed-batch fermentation process", *Transactions of the Institute of Measurement and Control*, 20(1), 11-18.

Chapter 3

The Inheritance and Generalizability Properties Extended from Function Definitions into Functionals

Abstract

According to the characteristic of Functional, that is a type of function we investigate the Inheritance and Generalizability of the function properties on functionals. Particularly, based on these properties, we show that on Euclidean spaces the new types of functionals can be presented as the generalization of their same-name functions.

Keywords: subset principle, axiom of induction, inheritance property, generalizability property, function, functional

3.1. Introduction

Functional analysis is a branch of mathematics concerned with infinite-dimensional vector spaces (mainly function spaces) and mappings between them. These mappings are called functionals if the range is on the real line. This research studies the properties of functionals based on the framework of functions. To do this, first, some basic information about functions shall be put forward from abstract analysis and topology by mentioning some definitions and theorems which this research shall need to use. Based on these, the research shall analyze the Inheritance and Generalizability properties in functionals. The final goals of the research which shall be investigated are:

1. To examine the existence of Inheritance and Generalizability properties in functionals based on functions framework and describe AC, IPH and CAR functionals based on these properties (see Chapters 3 and 4).
2. To examine the capability of the function optimization algorithms to work with functionals directly as an inheritance property and usage of the CAM to optimize functionals directly based on this property (see Chapters 5 and 6).
3. To combine the CAM and DISOPE algorithm to generate the initial solution of CAM on OCP.

3.2. Preliminaries from Set Theory

Suppose $\mathcal{A}, \mathcal{B}, \mathcal{C}$ and $\mathcal{D}$ are sets. The ordered pair $(\mathcal{A}, \mathcal{B})$ is defined by the set $\{\{\mathcal{A}\}, \{\mathcal{A}, \mathcal{B}\}\}$. Furthermore, the ordered triple $(\mathcal{A}, \mathcal{B}, \mathcal{C})$ is $((\mathcal{A}, \mathcal{B}), \mathcal{C})$ and the ordered quadruple $(\mathcal{A}, \mathcal{B}, \mathcal{C}, \mathcal{D})$ is $((\mathcal{A}, \mathcal{B}, \mathcal{C}), \mathcal{D})$.

In addition, a set is called a relation if and only if all its members are ordered pairs. Also, a set f is called a function if and only if f is a relation and for each $x \in domain(f)$ there exists a unique set $\mathcal{Y}$ such that $\mathcal{X}, \mathcal{Y} \in f(x, y)$, (Searcoid, 1948).

Traditionally a functional is a map from a vector space of functions usually to real numbers. In other words, it is a function that takes functions as its argument or input and returns a real number.

Furthermore, the following definitions are useful for establishing the Theorem 3.3 in this chapter.

Definition 3.1: A function $f: \mathcal{A} \rightarrow \mathcal{B}$ is said to be injective (or one-to-one) if for each pair of distinct points of $\mathcal{A}$, their image under f are distinct [1].

Definition 3.2: It is said to be surjective (or f is said to map $\mathcal{A}$ onto $\mathcal{B}$) if every element of $\mathcal{B}$ is the image of some element of $\mathcal{A}$ under the function f.

Definition 3.3: If f is both injective and surjective, it is said to be bijective or a one-to-one correspondence.

Remark 3.1: Formally, f is injective if $[f(a) = f(a')] \Rightarrow [a = a']$, and f is surjective if $[\mathscr{b} \in \mathcal{B}] \Rightarrow [\mathscr{b} = f(a)$ for at least one $a \in \mathcal{A}]$.

Definition 3.4: A set is said to be finite if there is a bijective correspondence of $\mathcal{A}$ with some segment of the positive integers. That is, $\mathcal{A}$ is finite if it is empty or if there is a bijection $f: \mathcal{A} \rightarrow \{1, \ldots, n\}$.

Remark 3.2: A set $\mathcal{A}$ is said to be infinite if it is not finite.

Definition 3.5: A set $\mathcal{A}$ is said to be countably infinite if there is a bijective correspondence $f: \mathcal{A} \rightarrow \mathbb{Z}_+$.

Remark 3.3: A set $\mathcal{A}$ is said to be countable if it is either finite or countably infinite. A set that is not countable is said to be uncountable.

Let $\mathcal{A}$ be a set of all functions. A map is described from the set of natural numbers to set $\mathcal{A}$ which implies that set $\mathcal{A}$ is countable. The following theorems and corollary are crucial to the discussion.

Theorem 3.1: *Let $\mathcal{A}$ be a nonempty set. Then the following are equivalent*:

(a) *A is countable.*
(b) *There is a surjective function* $f: \mathbb{Z}_+ \longrightarrow \mathcal{A}$.
(c) *There is an injective function* $g: \mathcal{A} \longrightarrow \mathbb{Z}_+$.

Theorem 3.2: *A countable union of countable sets is countable.*

Corollary 3.1: A subset of a countable set is countable.

Definition 3.6 (Indexed Family): Let $\mathcal{F}$ be a nonempty collection of sets. An indexing function for $\mathcal{F}$ is a surjective function f from set $\mathcal{E}$, called the index set, to $\mathcal{F}$. The collection $\mathcal{F}$, together with the indexing function f, is called indexed family of sets. Given $d \in \mathcal{E}$, the set $f(d)$ shall be denoted by the symbol f_d. And the indexed family itself shall be denoted by the symbol$\{f_d\}_{d\in\mathcal{E}}$. Suppose that $f: \mathcal{E} \longrightarrow \mathcal{F}$ is an indexing function for $\mathcal{F}$; Let f_d denote $f(d)$. Then,

$$\bigcup_{d\in\mathcal{J}} f_d = \{x | for\ at\ least\ one\ d \in \mathcal{J}, x \in f_d\}.$$

Definition 3.7 (Axiom of Induction): The basic assumption of induction is, in logical symbols,

$$\forall \text{ predicates } \Theta, (\Theta(0) \wedge \forall k \in \mathbb{N}[\Theta(k) \Rightarrow \Theta(k+1)]) \Rightarrow \forall n \in \mathbb{N}\ \Theta(n)$$

where Θ is the proposition in question and k and n are both natural numbers. In other words, the basis $\Theta(0)$ being true along with the inductive case ("$\Theta(k)$ is true implies $\Theta(k+1)$ is true for all natural k") being true together imply that $\Theta(n)$ is true for any natural number n.

3.3. Generalizability Property of Function Concepts into Functionals Definitions

Based on Definitions 3.6 and 3.7, the procedure for constructing a model for the purpose of investigating the existence of Inheritance and Generalizability properties shall be presented. As a convention this model shall be called the Seyedi-Rohanin model (SRM):

Procedure 3.1

(I) Categorize set $\mathcal{A}$ into all their subset groups via their names, i.e., $\mathcal{B}_1$ be a set of all convex functions, $\mathcal{B}_2$ be a set of all linear functions, etc. Then index all these subset groups via indexed family (see Definition 3.6) i.e., f_{d_1} for the set $\mathcal{B}_1$, f_{d_2} for the set $\mathcal{B}_2$, etc. Then,

$$\alpha: f(d) = f_d,$$

$$\alpha: \mathbb{Z}_+ \longrightarrow \mathcal{A}$$

Based on Definition 3.3 and Remark 3.1:

$d \neq c \rightarrow f_d \neq f_c \rightarrow f(d) \neq f(c)$, implies α is injective and corange(α)= $\bigcup_1^\infty \mathcal{B}_m$= $\mathcal{A}$ implies that α are surjective, $\forall f_d \in \mathcal{A} \Rightarrow f_d = f(d), \exists\, d \in \mathbb{N}$. Thus, α is bijective.
The sets $\mathcal{B}_m$ are countably infinite ($m = 1, 2, \ldots$). $\mathcal{A} = \bigcup_1^\infty \mathcal{B}_m$ is countably infinite, because of $\bigcup_1^\infty \mathcal{B}_m$ is union of countably infinite sets (see Theorem 3.2). It implies that $\mathbb{N}$ and set $\mathcal{A}$ have a one-to-one correspondence. Thus, it implies that set $\mathcal{A}$ is countably infinite (see Theorem 3.1 and Figure 3.1).

(II) Similarly, set $\mathcal{Z}$ is categorized into all their subsets groups via their names, i.e., convex functionals, linear functionals, etc. Then all these groups are indexed (see Figure 3.2).

(III)Describe a map from set $\mathcal{A}$ to set $\mathcal{Z}$ to demonstrate their one-to-one correspondence. The following theorem shall establish the bijectivity of the map which was described between the set $\mathcal{A}$ and set $\mathcal{Z}$ in the model.

Theorem 3.3: *Let the set $\mathcal{A}$ be the set of all functions and set $\mathcal{Z}$ be the set of all functionals modelled via SRM. If (α, β) represents the bijective maps from the natural number to both sets, then ψ as a map from set $\mathcal{A}$ to $\mathcal{Z}$is bijective.*

Proof: By the assumption that α and β are bijective implies that

$$\begin{cases} \alpha \text{ is injective} \\ \beta \text{ is injective} \end{cases} \Rightarrow \begin{cases} i \neq e \rightarrow f_i \neq f_e \rightarrow: f(i) \neq: f(e) \\ i \neq e \rightarrow g_i \neq g_e \rightarrow: g(i) \neq: g(e) \end{cases} \tag{3.1}$$

and

$$\begin{cases} \alpha \text{ is surjective} \\ \beta \text{ is surjective} \end{cases} \Rightarrow \begin{cases} \forall f_i \in \mathcal{A} \Rightarrow \ f_i = f(i), \exists\, i \in \mathbb{N} \\ \forall g_i \in \mathcal{Z} \Rightarrow \ g(i) = g_i, \exists\, e \in \mathbb{N} \end{cases}. \tag{3.2}$$

Based on (3.1), (3.2), Definition 3.3 and Remark 3.1:

$$\begin{cases} \psi: \mathcal{A} \longrightarrow \mathcal{Z}: i \neq e \ \rightarrow f_i \neq \ g_e \ \rightarrow f(i) \neq g(e) \\ \psi: \mathcal{A} \longrightarrow \mathcal{Z}: \forall g_i \in \mathcal{Z} \ \Rightarrow \ \exists\, f_i \in \mathcal{A}: f_i = g_i \end{cases} \begin{cases} \Rightarrow \psi \textit{ is injective} \\ \Rightarrow \psi \textit{ is surjective} \end{cases}$$

Then, ψ is bijective (see Figure 3.3) ∎.

(IV)Let Θ be a predicate that performs the existence of the bijective maps $\psi_{k}: \mathcal{A} \longrightarrow \mathcal{Z}$, which connect the similar elements of the sets together (where $\mathcal{k} \in \mathbb{N}$). Based on the construction of the sets $\mathcal{A}$ and $\mathcal{Z}$, it is clear while the sets $\mathcal{A}$ and $\mathcal{Z}$ are empty then $\Theta(0)$ is true, because it is also empty. And while the sets $\mathcal{A}$ and $\mathcal{Z}$ have one element i.e., $\mathcal{B}_1$ (set of all convex functions) and $\mathcal{B}_1^*$ (set of all convex functionals) then $\Theta(1)$ is true because there is a bijective map ψ_1 which connects both of them together based on Definition 4.6 and Theorem 3.3. Also while the sets $\mathcal{A}$ and $\mathcal{Z}$ have two elements i.e., $\mathcal{B}_1$ (set of all convex functions), $\mathcal{B}_2$ (set of all linear functions) and $\mathcal{B}_1^*$ (set of all convex functionals), $\mathcal{B}_2^*$ (set of all linear functionals) then $\Theta(2)$ is true because there are two bijective maps ψ_1 and ψ_2 which connect $\mathcal{B}_1$ to $\mathcal{B}_1^*$ and $\mathcal{B}_2$ to $\mathcal{B}_2^*$ respectively based on Definition 3.6 and Theorem 3.3. It means that $\Theta(2)$ is true because $\Theta(1)$ is true. By the continuity of the process, it is also clear that $\Theta(3)$ is true because $\Theta(2)$ is true. Finally, $\Theta(\mathcal{k})$ is true because $\Theta(\mathcal{k} - 1)$ is true and also $\Theta(\mathcal{k} + 1)$ is true because $\Theta(\mathcal{k})$ is true. It induces that the conditions of Definitions 3.7 hold and all elements of the Set $\mathcal{A}$ is connected to their similar elements of the Set $\mathcal{Z}$ which have the same name i.e., $\mathcal{B}_{\mathcal{k}}$ (affine functions) and $\mathcal{B}_{\mathcal{k}}^*$ (affine functionals), (see Figure 3.3).

As a result of the above procedure, the following remark will appear.

Remark 3.1 (Generalizability Property): SRM demonstrates that if there is a type of function then there is its same-name functional (both defined in non-dual space) i.e., if there is a linear function then there is linear functional (see Figure 3.3).

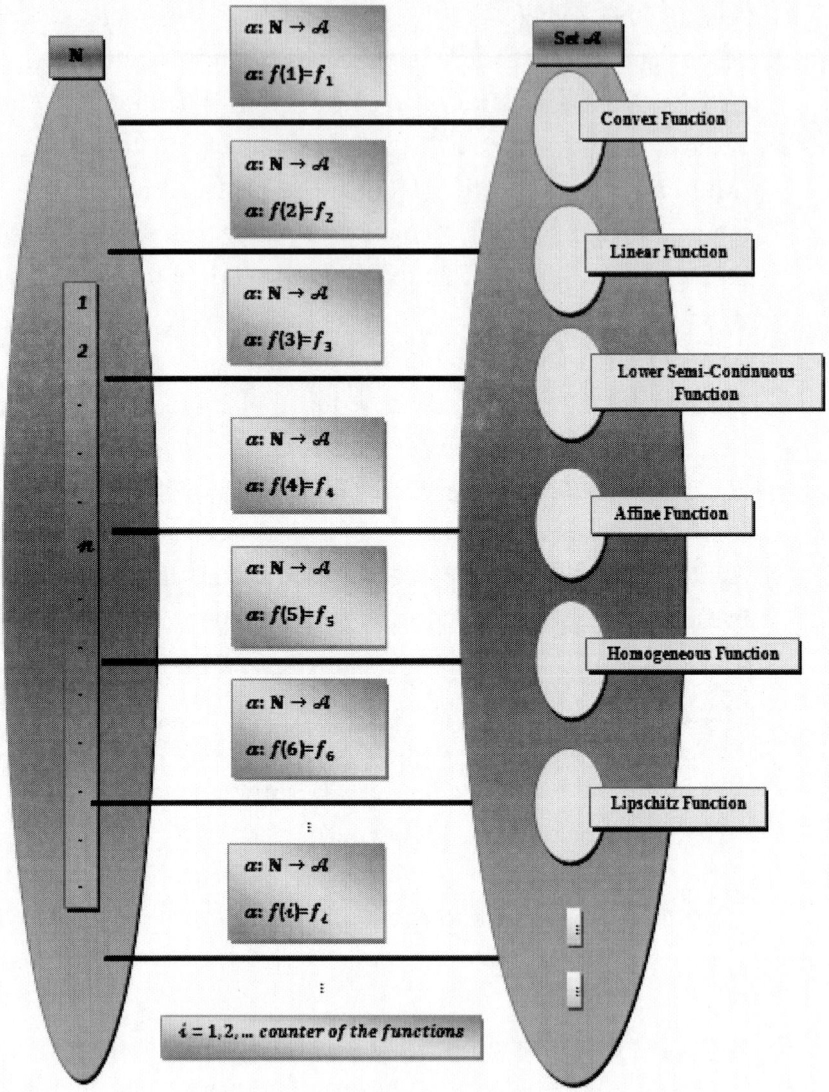

Figure 3.1. Existence of bijective maps from Numbers set to countably set of all functions via SRM.

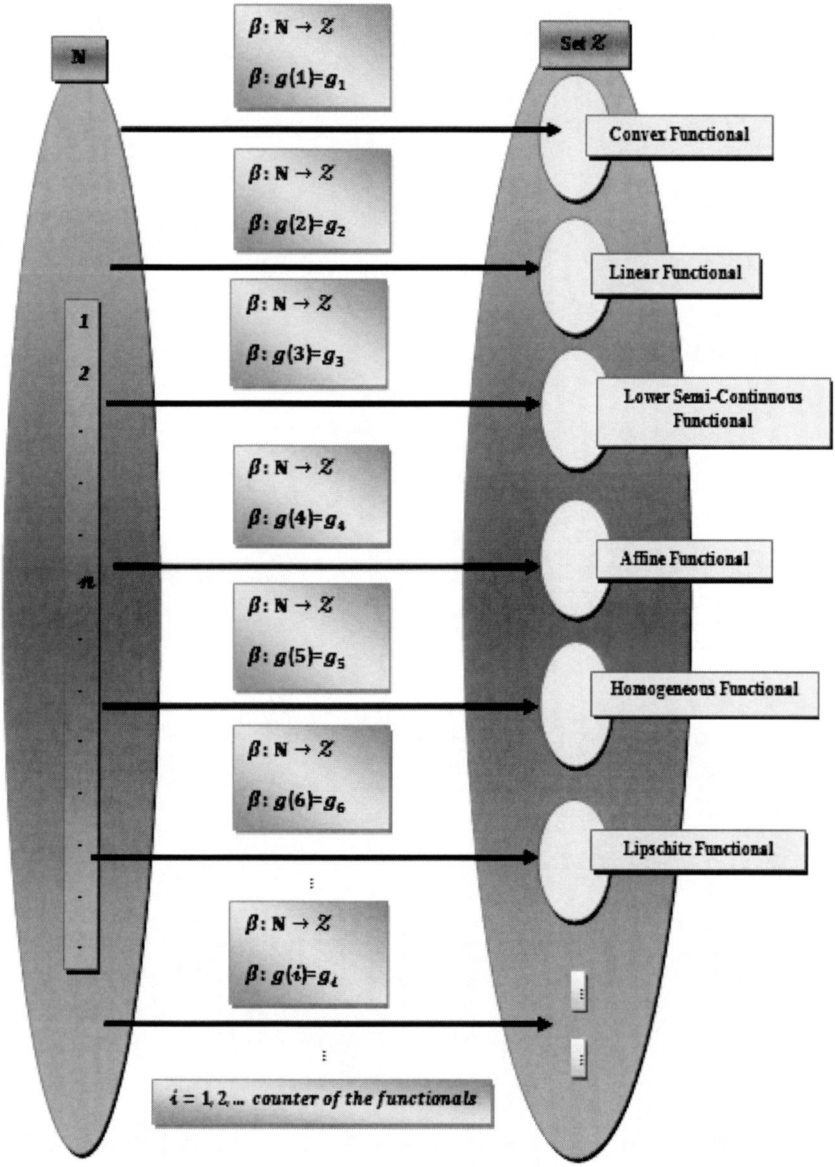

Figure 3.2. Existence of bijective maps from the natural number set to countably set of all functionals via SRM.

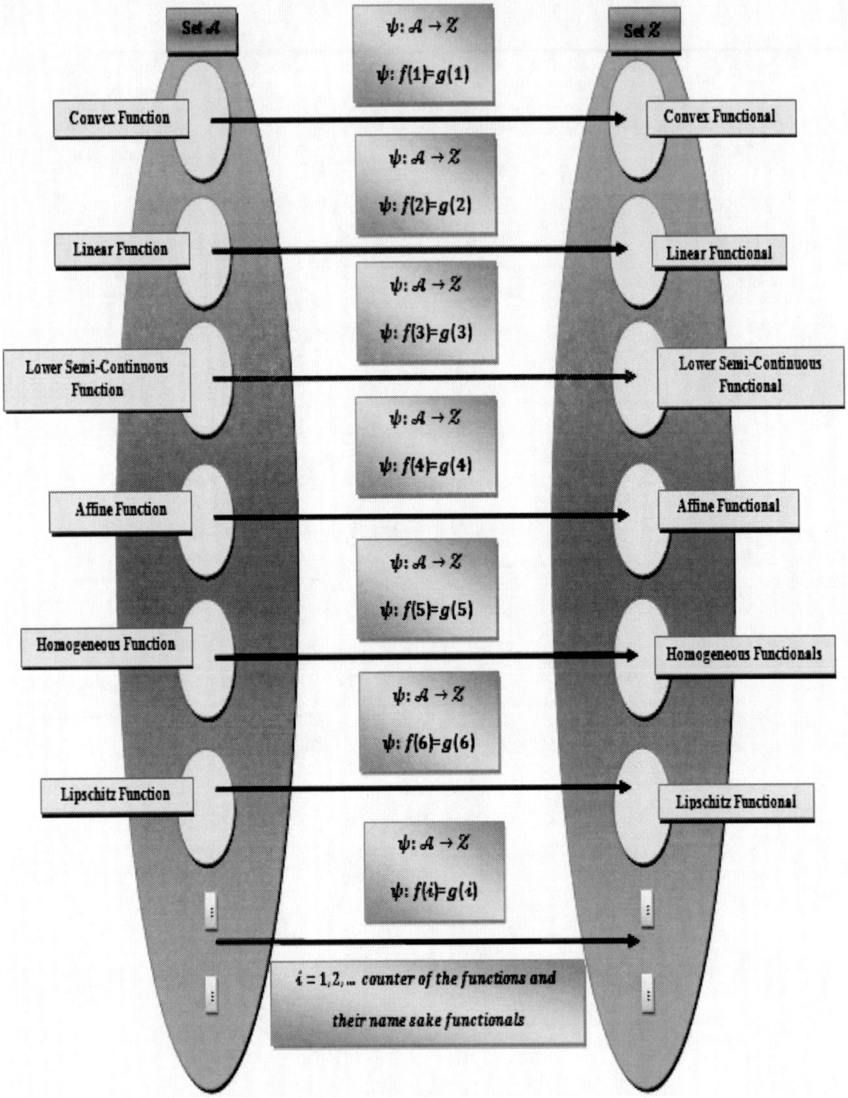

Figure 3.3. Existence of bijective maps from the set of all functions to the set of all functionals via SRM.

3.4. Inheritance of Function Property in the Structure of Functionals

In this section, the inheritance properties of the function are described based on the following theorem. Furthermore, it is clear that $\mathcal{Z}$ (set of all functionals) is a subset of $\mathcal{A}$ (set of all functions). It means that according to the definition of functional where it is a type of function then $\mathcal{Z} \subseteq \mathcal{A}$.

Theorem 3.4 (Subset Principle): *If $\Xi(x)$ is a condition on sets, then, for each set $\mathcal{A}$, there exists a unique set whose members are precisely those members $\mathcal{Z}$ of $\mathcal{A}$ for which $\Xi(\mathcal{Z})$ holds. where $\Xi(.)$ is the Laws and Properties of functions.*

Remark 3.2 (Inheritance Property): According to Theorem 3.4, it is clear that subsets inherit their properties from their superset. It implies that any functional should observe all function laws and properties as its inheritance property.

SRM, Theorem 3.4, Theorem 3.3 and bijectivity of the map ψ demonstrate that there exist the following results, such that they confirm the existence of Inheritance and Generalizability properties from functions in functionals:

1. All the functionals inherited their properties from their same-name functions.
2. All the functionals are generalization of their same-name functions
3. if there is a type of function then there is its same-name functional i.e., if there is a concave function then there is concave functional

As a result of the above discussion, the SRM proposes the following theorem based on the Inheritance and Generalizability properties generally.

Theorem 3.5: *If and only if there is a definition for functions then there is its corresponding definition for functionals.*

Proof: Based on Procedure 1:

1. Similarly, describe two sets $\mathcal{U}_1$ as the set of all function definitions which is countable and indexed via SRM and $\mathcal{U}_2$ as the set of all functional definitions which is countable and indexed via SRM. The proof is alike to previous action completely (see Figure 3.4).
2. Also, let Θ be the statement that performs the existence of the bijective maps $\psi_{\mathcal{k}}: \mathcal{U}_1 \longrightarrow \mathcal{U}_2$, which connect the similar elements of the sets together (where $\mathcal{k} \in \mathbb{N}$). Based on construction of the sets $\mathcal{U}_1$ and $\mathcal{U}_2$, it is clear while the sets $\mathcal{U}_1$ and $\mathcal{U}_2$ have one element i.e., Δ_1 (Abstract convexity definition for functions) and $\overline{\overline{\Delta_1}}$ (Abstract convexity definition for functionals) then $\Theta(1)$ is true because there is a bijective map ψ_1 which connects both of them together based on Theorem 3.3. Also while the sets $\mathcal{U}_1$ and $\mathcal{U}_2$ have two elements i.e., Δ_1 (Abstract convexity definition for functions), Δ_2 (Homogeneity definition for functions) and $\overline{\overline{\Delta_1}}$ (Abstract convexity definition for functionals), $\overline{\overline{\Delta_2}}$ (Homogeneity definition for functionals) then $\Theta(2)$ is true because there are two bijective maps ψ_1 and ψ_2 which connect Δ_1 to $\overline{\overline{\Delta_1}}$ and Δ_2 to $\overline{\overline{\Delta_2}}$ respectively based on Theorem 3.3. It means that $\Theta(2)$ is true because $\Theta(1)$ is true. By the continuous, it is also clear that $\Theta(3)$ is true because $\Theta(2)$ is true, etc. Finally, $\Theta(\mathcal{k})$ is true because $\Theta(\mathcal{k}-1)$ is true and also $\Theta(\mathcal{k}+1)$ is true because $\Theta(\mathcal{k})$ is true. It induces that the conditions of Definitions 3.7 hold and all elements of the Set

$\mathcal{U}_1$ is connected to their similar elements of the Set $\mathcal{U}_2$ which have the same name i.e., $\Delta_{\mathscr{k}}$ (affinity definition for functions) and $\overline{\overline{\Delta_{\mathscr{k}}}}$ (affinity definition for functionals).

Inheritance and Generalizability properties and Theorem 3.5 claimed that all the definitions of functions can be extended to introduce new definitions on functionals i.e., convexity-along-ray definition on function can be extended to introduce this definition for functional. Next section shall show some well-known examples to investigate the validity of the Theorem 3.5 experimentally.

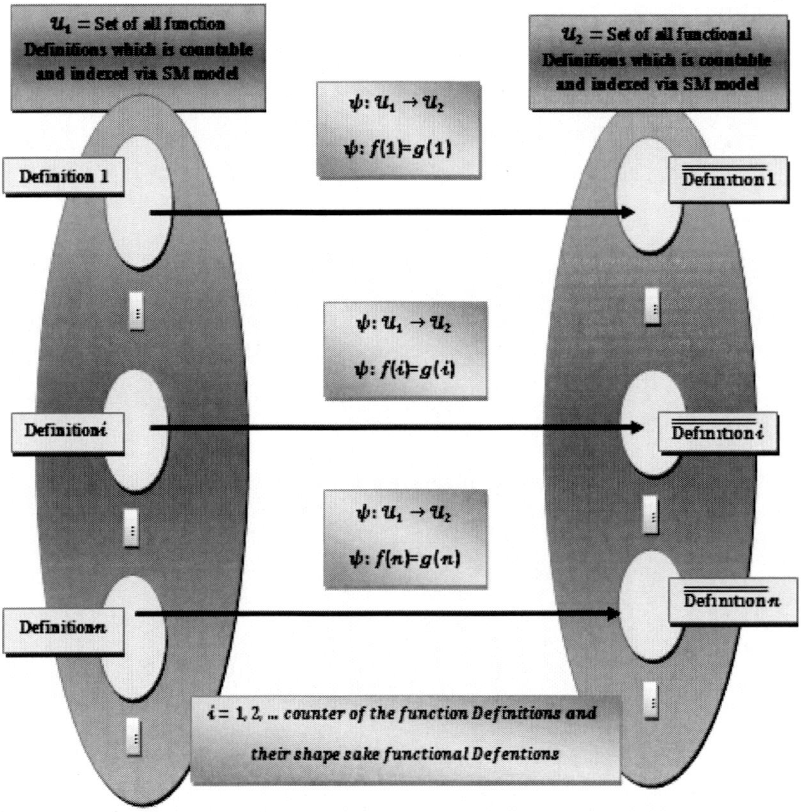

Figure 3.4. Existence of bijective maps from the set of all function Definitions to the set of all functional Definitions via SRM.

3.5. Several Examples of the Inheritance and Generalizability Properties of Function Definitions in the Body of the Functionals

This part shall demonstrate the validity of the Remark 3.1 and Remark 3.2 via some important examples from functional analysis.

3.5.1. Convexity in Functionals

Before the convexity of functional is discussed, a relevant concept in functions is first mentioned.

Definition 3.8 (Convex Function): Convexity of a real-valued function f on an interval $\mathcal{I}$ is often defined by the condition that

$$f((1-\mathscr{b})x + \mathscr{b}y) \leq (1-\mathscr{b})f(x) + \mathscr{b}f(y)$$

whenever $x, y \in \mathcal{I}$ and $0 \leq \mathscr{b} \leq 1$. [2]. Based on this, the discussion on convexity of functionals shall be shown as follow.

Definition 3.9 (Convex Functional): A functional, defined on a convex subset of a linear vector space, is the supergraph of which is a convex set. A functional F which does not assume the value $-\infty$ on a convex set $\mathcal{A}$ is convex on $\mathcal{A}$ if and only if the inequality

$$F((1-\mathscr{b})x + \mathscr{b}y) \leq (1-\mathscr{b})F(x) + F(y), x, y \in \mathcal{A}, 0 \leq \mathscr{b} \leq 1,$$

is satisfied [3].

Next, based on the above definition of the function the next definition is stated.

3.5.2. Continuity in Functionals

The continuity of a functional is another property which is important in this study. The definition of a continuous functional will be based on the following definition of a continuous function.

Definition 3.10 (Continuous Function): Let f be a real-valued function defined on a subset $\mathbb{E}$ of the real numbers $\mathbb{R}$, that is, $f: \mathbb{E} \to \mathbb{R}$. Then f is said to be continuous at a $x_0 \in \mathbb{E}$ if for any $\varepsilon > 0$ there exists a $\delta > 0$ such that for all $x \in \mathbb{E}$ with $|x - x_0| < \delta$ the inequality $|f(x) - f(x_0)| < \varepsilon$ is valid [4].

Based on the above property of the function, the next definition is stated.

Definition 3.11 (Continuous Functional): A functional $F: \mathcal{M} \to \mathbb{R}$, where $\mathcal{M}$ is a subset of a topological space $\breve{\mathcal{X}}$, to be continuous at a vector $x_0 \in \mathcal{M}$, for any $\varepsilon > 0$ there must be a neighbourhood $\mathcal{U}$ of x_0 such that $|F(x) - F(x_0)| < \varepsilon$ for $x \in \mathcal{U}$ (Sobolev, 2001).

3.5.3. Lower Semi-Continuity in Functionals

The discussion now proceeds to lower semi-continuity of functionals. As before similar concept in functions is first presented.

Definition 3.12 (Lower Semi-Continuous Function): A real-valued function $f(x)$ is lower semi-continuous at a vector x_0 if, for any small positive number ε, $f(x)$ is always greater than $f(x_0) - \varepsilon$ for all x in some neighborhood of x_0.

Based on Definition 3.12 the following definition is introduced.

Definition 3.13 (Lower Semi-Continuous Functional): Let $\breve{\mathcal{X}}$ be a metric space. A functional $F: \breve{\mathcal{X}} \to \overline{\mathbb{R}}$ ($\overline{\mathbb{R}} = \mathbb{R} \cup \{+\infty, -\infty\}$) is lower semicontinuous at a vector $x_0 \in \mathcal{X}$ if [5].

$$F(x_0) \leq \lim_{x \to x_0} Inf\ F(x)$$

3.5.4. Linearity in Functionals

In this section, a discussion of linearity will be in order.

Definition 3.14 (Linear Function): A function $l: \mathbb{R}^m \to \mathbb{R}^n$ is linear if for any vectors $\tilde{x}$ and $\tilde{y}$ in $\mathbb{R}^m$,

$$l(\tilde{x} + \tilde{y}) = l(\tilde{x}) + l(\tilde{y}),$$

and for any vector $\tilde{x}$ in $\mathbb{R}^m$ and scalar $\mathfrak{s}$, $l(\mathfrak{s}\tilde{x}) = \mathfrak{s}l(\tilde{x})$, [6].

Based on the Definition 3.14, the following definition of a linear functional is presented.

Definition 3.15 (Linear Functional): A linear functional on a real vector space $\breve{\mathcal{V}}$ is a function $t : \breve{\mathcal{V}} \to \mathbb{R}$, which satisfies the following properties.

1. $t(\mathcal{v} + \mathcal{w}) = t(\mathcal{v}) + t(\mathcal{w})$, and
2. $t(\mathfrak{n}\mathcal{v}) = \mathfrak{n}t(\mathcal{v})$.

when $\breve{\mathcal{V}}$ is a complex vector space and $\mathfrak{n}$ is a real number, then t is a linear map onto the complex numbers. Generalized functions are a special case of linear functionals, and have a rich theory surrounding them [7].

3.5.5. Affinity in Functionals

The discussion now proceeds to affinity, starting with the definition of affinity in functions.

Definition 3.16 (Affine Function): A function $g: \mathbb{R}^m \to \mathbb{R}^n$ is affine if there is a linear function $l: \mathbb{R}^m \to \mathbb{R}^n$ and a vector $\tilde{b}$ in $\mathbb{R}^n$ such that

$$g(\tilde{x}) = l(\tilde{x}) + \tilde{b}$$

for all $\tilde{x}$ in $\mathbb{R}^m$ [6].

Based on Definition 2.10 the definition of affine functional is introduced next.

Definition 3.17 (Affine Functional): Let $\breve{\mathcal{X}}$ be a linear space over the real number field, a functional G is affine on $\breve{\mathcal{X}}$ if

$$G(x) = F(x) + d$$

where F is a linear functional on $\breve{\mathcal{X}}$ and d is a real number [8].

3.5.6. Homogeneity in Functionals

This section starts with the definition of a homogeneous function.

Definition 3.18 (Homogeneous Function): Suppose $\breve{\mathcal{V}}, \breve{\mathcal{W}}$ are vector spaces over $\mathbb{R}$, and $f: \breve{\mathcal{V}} \to \breve{\mathcal{W}}$ is a mapping.

- If there exists a $r \in \mathbb{R}$, such that

  $$f(bv) = b^r f(v)$$

  for all $b \in \mathbb{R}$ and $v \in \mathcal{V}$, then f is a homogeneous function of degree r.
- If there exists a $r \in \mathbb{R}$, such that

  $$f(bv) = b^r f(v)$$

  for all $b \geq 0$ and $v \in \mathcal{V}$, then f is a positively homogeneous function of degree r [4].

Based on the above property of functions, the homogeneity of a functional is defined as follows.

Definition 3.19 (Homogeneous Functional): Let $\breve{\mathcal{X}}$ is a vector space and $\mathcal{K}$ is a convex cone in $\breve{\mathcal{X}}$. A functional $F:\mathcal{K} \to \mathbb{R}$ is homogeneous if $F(\mathfrak{m}x) = \mathfrak{m}F(x)$ for all $x \in \mathcal{K}$ and all $\mathfrak{m} \in \mathbb{R}$ (and a homogeneous functional F is positively homogeneous if $\mathfrak{m} \geq 0$), [9].

REFERENCES

[1] Munkres, J. R. (2000). "*Topology Prentice Hall*", Upper Saddle River, NJ.

[2] Barbu, V. & Precupanu, T. (2012). *Convexity and optimization in Banach spaces*: Springer.

[3] Roekafellar, R. (1970). "*Convex analysis*", Princeton.

[4] Kudryavtsev, L. (2002). "*Homogeneous function", Encyclopaedia of mathematics*. Springer, Berlin.

[5] Kurdila, A. J. & Zabarankin, M. (2005). "*Convex Functional Analysis, Systems and Control: Foundations and Applications*," ed: Birkhäuser, Basel.

[6] Sloughter, D. (2001). "*The Calculus of Functions of Several Variables,*" ed: Furman University, 260p.

[7] Boyd, S. P., El Ghaoui, L., Feron, E. & Balakrishnan, V. (1994). *Linear matrix inequalities in system and control theory*, vol. 15, SIAM.

[8] Bittner, L. (1974). "On moment equations and inequalities", *Lithuanian Mathematical Journal*, 14(2), 178-185.

[9] Castagnoli, E. & Maccheroni, F. (2000). "Restricting independence to convex cones", *Journal of Mathematical Economics*, 34(2), 215-223.

Chapter 4

STUDY OF SOME NEW TYPE OF FUNCTIONALS DEFINED BASED THE INHERITANCE AND GENERALIZABILITY PROPERTIES OF FUNCTIONS

ABSTRACT

As a major branch of Euclidean geometry, convexity has its proof in the field of algebra and analysis. This study investigates increasing positively homogenous; abstract convex; convex-along-rays functionals defined in abstract convexity. In particular, based on the Inheritance and Generalizability of function properties on functionals, we analyze the existence of subdifferentiability of these functionals on R^n.

Keywords: positive homogeneity, abstract convexity, convexity-along-ray, subdifferential, functional

4.1. INTRODUCTION

The main results of this chapter are existence of the Inheritance and Generalizability properties in both functions and functionals. So, the

Generalizability property supports that if there is any type of function (defined on non-dual space) then there is its same-name functional as its extension. Also, inheritance property also confirms that the properties of function are inherited in its same-name functional inherently. The major tool which demonstrates these properties is SRM. This research proposed SRM based on indexed family.

This chapter shall propose AC, CAR and IPH functionals based on their same-names functions and Theorem 3.5. So, the Generalizability property supports that if there are AC, CAR and IPH functions then there are AC, CAR and IPH functionals. Also, Inheritance property confirms that the properties of AC, CAR and IPH functions are inherited in their same-name functionals inherently. The three subsections shall show these definitions via the properties.

4.2. Increasing Positively Homogeneous Functionals on the Euclidean Cone

This part is going to examine some classes of monotonic functionals defined on either the cone $\mathbb{R}^n_+ = \{x \in \mathbb{R}^n : x_i \geq 0 \; for \; all \; i = 1, \dots, n\}$ or the cone $\mathbb{R}^n_{++} = \{x \in \mathbb{R}^n : x_i > 0 \; for \; all \; i = 1, \dots, n\}$ and the so-called normal subsets of these cones. And these classes shall be investigated in the abstract convexity field. A nonempty set $\mathcal{U} \in \mathbb{R}^n_+$ is said to be normal if

$$(x \in \mathcal{U}, x' \in \mathbb{R}^n_+, x' \leq x) \Rightarrow x' \in \mathcal{U}$$

In this research, let the empty set be a normal subset of $\mathbb{R}^n_+$. Normal sets are connected to the so-called increasing positively homogeneous of degree one (IPH) functionals. There is an obvious similarity between the class of IPH functionals and the class of sublinear functionals, and between normal sets and convex sets. One of the major tools for the study of Sublinearity of functionals and the convexity of sets are linear functionals.

For instance, a functional P defined on $\mathbb{R}^n$ is lower semicontinuous and sublinear if and only if this functional is abstract convex which shall be discussed next with respect to the set of all linear functionals, on the other hand, there exists a set of linear functionals U such that $P(x) = \sup\{u(x): u \in \mathcal{U}\}$.

4.3. Preliminaries from Convex Analysis

This section is going to put forward some related definitions and properties which Anger (1977), Rubinov (2000), Nowak and Trybulec, (2004), Hamel (2004) and Kasyanov et al. (2008) stated in their papers for functions and functionals.

Firstly, let $F: \tilde{Q} \to \mathbb{R}$ be a functional and $\tilde{Q}$ be real linear space.

1. F is subadditive if and only if:
   for all vectors x, y of $\tilde{Q}$ holds $F(x + y) \leq F(x) + F(y)$, [1].
2. F is additive if and only if:
   for all vectors x, y of $\tilde{Q}$ holds $F(x + y) = F(x) + F(y)$,.
3. F is homogeneous if and only if:
   for every vector x of $\tilde{Q}$ and for every real number b holds $F(b.x) = b.F(x)$,.
4. F is positively homogeneous if and only if:
   for every vector x of $\tilde{Q}$ and for every real number b such that $b > 0$ holds $F(b.x) = b.F(x)$,.
5. F is semi-homogeneous if and only if:
   for every vector x of $\tilde{Q}$ and for every real number b such that $b \geq 0$ holds $F(b.x) = b.F(x)$, [2].

6. F is absolutely homogeneous if and only if:
   for every vector x of $\tilde{Q}$ and for every real number b holds $F(b.x) = |b|.F(x)$, [3].

Secondly, let us consider $\tilde{Q}$. The following observations stated by Popiolek (1989):

(a) every functional in $\tilde{Q}$ which is additive is also subadditive.
(b) every functional in $\tilde{Q}$ which is homogeneous is also positively homogeneous.
(c) every functional in $\tilde{Q}$ which is semi-homogeneous is also positively homogeneous.
(d) A linear functional in $\tilde{Q}$ is an additive homogeneous functional in $\tilde{Q}$.

Additionally, the following propositions stated by Nowak and Trybulec, (2004) and Hamel (2004):

(a) For every functional G in $\tilde{Q}$ and for every vector v of $\tilde{Q}$ holds $G(-v) = -G(v)$, (Nowak and Trybulec, 2004).
(b) For every functional G in $\tilde{Q}$ and for every vectors v_1, v_2 of $\tilde{Q}$ holds $G(v_1 - v_2) = G(v_1) - G(v_2)$ [4].
(c) It is well-known that a positively homogeneous functional is convex if and only if it is subadditive [5].
(d) A positive homogeneous and subadditive function is called Sublinear.

The next part shall propose new functionals based on the above definitions and propositions.

4.4. Increasing Positively Homogeneous Functional Definition on Euclidean Space

The definition of IPH functional shall be put forward and extended as stated by Anger (1977).

Let $\tilde{Q}$ be either $\mathbb{R}_+^n$ or $\mathbb{R}_{++}^n$. A functional F defined on $\tilde{Q}$ is called increasing if $x \geq y$ implies that $F(x) \geq F(y)$. The functional F is positively homogeneous of degree one if $F(bx) = bF(x)$ for all $x \in \tilde{Q}$, $b > 0$ and $b \in \mathbb{R}_+$. By combining these definitions the following definition of an IPH functional is obtained.

Definition 4.1: Let $\tilde{Q}$ be either $\mathbb{R}_+^n$ or $\mathbb{R}_{++}^n$. A functional $F: \tilde{Q} \to \mathbb{R}_{+\infty}$ is an IPH functional if the following conditions are satisfied:

1. $x \geq y$ implies that $F(x) \geq F(y)$:
2. $(bx) = bF(x)$ for all $x \in \tilde{Q}$ and $b > 0$ and $b \in \mathbb{R}_+$, [6].

Now some examples of IPH functionals are given.

1. Let $\tilde{\mathcal{E}}$ be a locally compact Hausdorff space and let $\mathcal{R}$ be the set of all compact subsets of $\tilde{\mathcal{E}}$. Let $\mathcal{K}$ (respectively $\mathcal{K}_+$) denote the set of positive real-valued upper semicontinuous (respectively continuous) functions on $\tilde{\mathcal{E}}$ with compact support. And $c: \mathcal{R} \to \mathbb{R}_+$ is a positive and real-valued function defined on the compact subsets of $\tilde{\mathcal{E}}$. Then, the Capacity functionals in the form $C: h \to \int hdc$ which were stated by Anger (1997) is IPH functionals where $h \in \mathcal{K}_+$ and $\int hdc \in \mathbb{R}_+$.
2. Let $\widetilde{\mathcal{X}}$ be a topological space. Let $G: \widetilde{\mathcal{X}} \to \mathbb{R} \cup \{+\infty\} \cup \{-\infty\}$ be a functional. Hamel (2004) defined the set $\mathcal{A}_G := \{x \in \widetilde{\mathcal{X}}: G(x) < 0\}$. Let $\mathcal{A} \subset \widetilde{\mathcal{X}}$ be a set and $k^0 \in \widetilde{\mathcal{X}}$. Then the functional $\begin{cases} G_{\mathcal{A}}(x) = \inf\{t \in \mathbb{R}: x + tk^0 \in \mathcal{A}\} \\ \text{and } x_1 \leq x_2 \Rightarrow G_{\mathcal{A}}(x_1) \leq G_{\mathcal{A}}(x_2) \end{cases}$ which was stated by Hamel (2004) is IPH functional.

4.5. Subdifferentialability of the Increasing Positively Homogeneous Functionals on the Euclidean Cone

The main definition of subdifferential which Kasyanov et al. (2008) stated on functional shall be put forward and the definition of subdifferential that Rubinov (2000) introduced for function shall be extended for use in explaining the subdifferential of IPH functionals.

Let $\widetilde{\mathcal{D}}$ be Frechet space, $\widetilde{\mathcal{D}}^*$ its topologically dual (adjoint) space. For $x \in \widetilde{\mathcal{D}}$ and $F \in \widetilde{\mathcal{D}}^*$. The symbol $\langle F, x\rangle$ show the bilinear parting between $\widetilde{\mathcal{D}}$ and $\widetilde{\mathcal{D}}^*$. Also, let the IPH functional $F: \widetilde{\mathcal{D}} \to \mathbb{R} \cup \{+\infty\}$ and the symbol $dom\ F$ denotes the set $\{x \in \widetilde{\mathcal{D}} | F(x) < +\infty\}$. Given an IPH functional F and a convex cone $\mathcal{U}$ such that $int\ \mathcal{U} \subset dom\ F$, a local subdifferential of F at the vector $x_0 \in \mathcal{U} \cap dom\ F$ is the set:

$$\partial F(x_0, \mathcal{U}) = \{v \in \widetilde{\mathcal{D}}^* | \langle v, x - x_0\rangle_{\widetilde{\mathcal{D}}} \leq F(x) - F(x_0) \text{ for all } x \in \mathcal{U}\}$$

Besides that, if $\mathcal{U}_1 \subset \mathcal{U}_2$ then $\partial F(x_0, \mathcal{U}_1) \subset \partial F(x_0, \mathcal{U}_2)$. Especially, $\partial F(x_0, \widetilde{\mathcal{D}}) = \partial F(x_0) \subset \partial F(x_0, \mathcal{U})$. The above set is called the subdifferential of F at the vector x_0.

By continuing of this process the definition of $\mathcal{L}$-subdifferential of IPH function which was stated by Rubinov (2000) for functional shall be extended, as follows.

Definition 5.2.: Let $\mathcal{L}$ be a set of simple functionals defined on a set $\mathcal{X}$. A functional $L \in \mathcal{L}$ is called an abstract subgradient (or $\mathcal{L}$-subgradient) of a proper functional $F: \mathcal{X} \to \mathbb{R}_{+\infty}$ at a vector y if $F(x) \geq L(x) - (L(y) - F(y))$ for all $x \in \mathcal{X}$. The set $\partial_{\mathcal{L}} F(y)$ of all abstract subgradient of F at y is referred to as the abstract subdifferential (or $\mathcal{L}$-subdifferential) of the functional F at the vector y.

Proposition 4.1: Let $\mathcal{L}$ be set of abstract linear functionals defined on a conic set $\mathcal{X}$ and let each $L \in \mathcal{L}$ be positively homogeneous of degree one ($L(\mathscr{b}x) = \mathscr{b}L(x)$ for all $\mathscr{b} > 0$). Then for an $\mathcal{L}$-convex functional P and for $\mathscr{y} \in \mathcal{X}$ thus:

$$\partial_{\mathcal{L}} P(\mathscr{y}) = \{L \in supp(P, \mathcal{L}) \colon L(\mathscr{y}) = P(\mathscr{y})\} \tag{4.1}$$

Proof: Similar to proof on functions by Rubinov (2000), let $L \in \partial_{\mathcal{L}} P(\mathscr{y})$, that is, $P(x) \geq L(x) - (L(\mathscr{y}) - P(\mathscr{y}))$ for every $x \in \mathcal{X}$. Let $x \in \mathcal{X}$. Investigate vectors $\mathscr{b}x$ with $\mathscr{b} > 0$. Then

$$\mathscr{b}P(x) = P(\mathscr{b}x) \geq \mathscr{b}L(x) - (L(\mathscr{y}) - P(\mathscr{y})) \tag{4.2}$$

$$\Rightarrow \mathscr{b}P(x) = L(x) - \frac{L(\mathscr{y}) - P(\mathscr{y})}{\mathscr{b}}.$$

Passing to the limit as $\mathscr{b} \to +\infty$, we get $P(x) \geq L(x)$, that is, $L \in supp(P, \mathcal{L})$. Specially, we have $L(\mathscr{y}) \leq P(\mathscr{y})$. On the other hand, passing to the limit in (4.2) as $\mathscr{b} \to +\infty$, we have $L(\mathscr{y}) \geq P(\mathscr{y})$. Therefore, $L(\mathscr{y}) = P(\mathscr{y})$ so $\partial_{\mathcal{L}} P(\mathscr{y})$ is contained in the set on the right-hand side in (4.1). Assume now that $P(x) \geq L(x)$ for all $x \in \mathcal{X}$ and $P(\mathscr{y}) - L(\mathscr{y}) = 0$. Then

$$P(x) \geq L(x) - (L(\mathscr{y}) - P(\mathscr{y})) \text{ for all } x \in \mathcal{X}.$$

Therefore the aim holds. ■

As mentioned before the $\mathcal{L}$-subdifferential of the functional P at a vector $\mathscr{y} \in \mathbb{R}^n_{++}$ is given by

$$\partial_{\mathcal{L}} P(\mathscr{y}) = \{L \in \mathcal{L} \colon P(x) \geq \langle L, x \rangle - \big(\langle L, \mathscr{y} \rangle - P(\mathscr{y})\big) \text{ for all } x \in \mathbb{R}^n_{++}\}.$$

Since the set $\mathcal{L}$ consist of positively homogeneous functionals, it exploits from the above proposition that

$$\partial_{\mathcal{L}} P(\mathscr{y}) = \{L \in supp(P, \mathcal{L}) \colon \langle L, \mathscr{y} \rangle = P(\mathscr{y})\}.$$

4.6. Abstract Convex Functional Defined on Euclidean Space

First of all, some beneficial definitions and theorems shall be put forward from functional analysis which will be needed for use in this section. Based on these, abstract convex functionals and their properties shall be defined. Our last aim is to state the subdifferential of abstract convex functionals. Let $\overline{\mathbb{R}} = \mathbb{R} \cup \{-\infty\} \cup \{+\infty\}$.

The following theorem will guide this research to reach its main aim that is definition of abstract convex functional.

Theorem 4.1 (*Hahn-Banach Theorem for Lower Semicontinuous Functionals*): *Let $\widetilde{\mathcal{X}}$ be normed vector space and suppose that the functional $F: \mathcal{X} \to \overline{\mathbb{R}}$, is proper, convex and lower semicontinuous. Then F is bounded below by an affine functional [7].*

The following definitions and theorems are extensions to propositions on functions by Rubinov, (2000). The next discussion on the convex-along-rays functionals will need these propositions

One of the main results of convex analysis exploited from the Theorem 4.1 states that an arbitrary lower semicontinuous convex functional F (perhaps admitting the value $+\infty$) is the upper envelop of the set of all its affine minorants:

$$F(x) = \sup\{H(x) \text{is an affine functional}, F(x) \leq F(x) \text{ for all } x. \quad (4.3)$$

The supremum of the above set is obtained at a vector x if and only if the subdifferential of F at x is nonempty. Since affine functionals are defined by means of linear functionals, one can say that convexity is linearity + envelope representation. In particular, abstract convex with respect to $\mathcal{H}$ or $\mathcal{H}$-convex functionals can be represented as upper envelopes of subsets of a set $\mathcal{H}$ in abstract convexity theory (convexity without linearity).

The set

$$\operatorname{supp}(F, \mathcal{H}) = \{H \in \mathcal{H}, H \leq F\}$$

of all $\mathcal{H}$-minorants of a functional F is called the support set (or $\mathcal{H}$-support set) of this functional. A set $\mathcal{U} \subset \mathcal{H}$ is called abstract convex or $(\mathcal{H}, \mathcal{X})$-convex if this set coincides with the support set of an abstract convex functional.

There are two equivalent definitions of the subdifferentials for a convex functionals. The first definition is based on the global behavior of the functional. A linear functional L is called a subgradient (i.e., a member of the subdifferential) of the functional F at a vector y if the affine functional

$$H(x) = L(x) - (L(y) - F(y))$$

is a support functional with respect to F, that is, $H(x) \leq F(x)$ for all x. On the other hand, the second definition has a local nature and related to a local approximation of the functional: the subdifferential is a closed convex set of linear functionals such that the directional derivative

$$F_x'(h) := \lim_{t \to +0} t^{-1} (F(x + th) - F(x))$$

is the upper envelope of this set. Both definitions illustrate the support and tangent sides of the gradient for a differentiable convex functional, respectively.

4.7. Preliminaries from Convex Functional Analysis and the Set Theory

The functionals $F: \mathcal{X} \to \mathcal{V}$ shall be studied where $\mathcal{X}$ is a set and $\mathcal{V} \subset \overline{\mathbb{R}}$ is a segment. Segment $\mathcal{V}$ as a rule will conform to one of the following segments:

$$\mathbb{R} = (-\infty, +\infty), \overline{\mathbb{R}} = [-\infty, +\infty], \mathbb{R}_{+\infty} = (-\infty, +\infty] \qquad (4.4)$$

and

$$\mathbb{R}_{+} = [0, +\infty), \overline{\mathbb{R}}_{+} = [0, +\infty], \mathbb{R}_{++} = (0, +\infty), \overline{\mathbb{R}}_{++} = (0, +\infty]. \qquad (4.5)$$

Let $\mathcal{V} \subset \overline{\mathbb{R}}$ be a segment. We shall investigate the supremum and the infimum of each subset of $\mathcal{V}$, specifically, supremum and the infimum of the empty subset $\emptyset$ of $\mathcal{V}$. It is appropriate to admit the following convention:

$$\inf \emptyset = \sup \mathcal{V}, \sup \emptyset = \inf \mathcal{V} \qquad (4.6)$$

Especially, $\inf \emptyset = +\infty$ for every $\mathcal{V}$ from (4.4) and (4.5). If $\mathcal{V}$ conform to one of the sets (4.4) then $\sup \emptyset = -\infty$; if $\mathcal{V}$ is one of the sets (4.5), then $\sup \emptyset = 0$. Let $\mathcal{V} \subset \overline{\mathbb{R}}$ be a segment and let $\mathcal{H}$ be a set of functionals $H: \mathcal{X} \to \mathcal{V}$. Assume that

$$\cup_{H \in \mathcal{H}} \{H(x) : x \in \mathcal{X}\} = \mathcal{V} \qquad (4.7)$$

The pointwise supremum and the pointwise infimum of each subset of $\mathcal{H}$ shall be investigated. In particular, the pointwise supremum and infimum of the empty subset, $\emptyset$, of $\mathcal{H}$ will be investigated. The following convention is conforming to (4.6):

$$\sup_{H \in \emptyset} H(x) = \inf \mathcal{V}, \inf_{H \in \emptyset} H(x) = \sup \mathcal{V}, x \in \mathcal{X} \qquad (4.8)$$

Therefore, the infimum of the empty subset of $\mathcal{H}$ is the constant functional equal to sup $\mathcal{V}$ and the supremum of the empty set of $\mathcal{H}$ is the constant functional equal to inf $\mathcal{V}$. Let $\mathcal{V} \subset \mathbb{R}$ be a segment and let $\bar{\mathcal{V}}$ be a closure of $\mathcal{V}$, that is,

$$\bar{\mathcal{V}} = \mathcal{V} \cup \{sup\ \mathcal{V}\} \cup \{inf\ \mathcal{V}\}. \tag{4.9}$$

Investigate a set $\mathcal{H}$ of functional $H: \mathcal{X} \to \mathcal{V}$. Let $\mathcal{U} \subset \mathcal{H}$ and $F_{\mathcal{U}}(x) = sup_{\mathrm{H} \in \mathcal{U}} H(x)$. If $\mathcal{U}$ is nonempty, then the functional $F_{\mathcal{U}}$ maps into $\mathcal{V} \cup \{sup\ \mathcal{V}\}$. If $\mathcal{U}$ is empty, then the functional $F(x) = inf\ \mathcal{V}$ for all $x \in \mathcal{X}$. Therefore, $F_{\mathcal{U}}: \mathcal{X} \to \bar{\mathcal{V}}$ for each $\mathcal{U} \subset \mathcal{H}$.

4.8. The Study of Some Properties of Abstract Convex Functionals

In this section, for describing the CAR functionals the main definitions related to abstract convexity which are extensions to propositions on functions by Rubinov (2000) shall be put forward.

Firstly, the following useful definition is explained.

Definition 4.2: Let $\mathcal{X}$ be a set, $\mathcal{V} \subset \overline{\mathbb{R}}$ and let $\mathcal{H}$ be a nonempty set of functionals $H: \mathcal{X} \to \mathcal{V}$. A functional $F: \mathcal{X} \to \bar{\mathcal{V}}$ is called abstract convex with respect to H (or H-convex) if there exists a set $\mathcal{U} \subset \mathcal{H}$ such that F is the upper envelop of this set:

$$F(x) = \sup\{H(x) \colon \mathrm{H} \in \mathcal{U}\}\ for\ all\ x \in \mathcal{X} \tag{4.10}$$

Let $\mathcal{L}$ be the set of all linear functionals defined on $\mathbb{R}^n$ and $\mathcal{H}$ be the set of all affine functionals defined on $\mathbb{R}^n$.

1.1. The following definition is based on affine operators [8] and the definition specifically to functionals is extended.

Definition 4.3: Let $\breve{\mathcal{X}}$ be a vector space and subset of $\mathbb{R}^n$. A functional $F: \breve{\mathcal{X}} \to \mathbb{R}$ is affine if there is a linear functional $L: \breve{\mathcal{X}}^{\#} \to \mathbb{R}$ and q an arbitrary element of $\breve{\mathcal{X}}$ such that

$$F(x) = L(x) + q \tag{4.11}$$

For all x in $\breve{\mathcal{X}}$. Where $\breve{\mathcal{X}}^{\#}$ is a uniquely determined vector subspace of $\breve{\mathcal{X}}$. On the other hand, an affine functional is just a linear functional plus a translation. However at this level we need to prove some theorems, in that case, first of all we are going to put forward the following definitions and theorems.

Both the following theorems proved by Yosida (1980) are useful for continuing this discussion:

Theorem 4.2: *Let $\breve{\mathcal{X}}$ be a real linear topological space, x_0 a vector of $\breve{\mathcal{X}}$ and $L(x)$ a continuous semi-norm on $\breve{\mathcal{X}}$. Then there exists a continuous linear functional F on $\breve{\mathcal{X}}$ such that $F(x_0) = L(x_0)$ and $|F(x)| \leq L(x)$ on $\breve{\mathcal{X}}$.*

Theorem 4.3: *Let $\breve{\mathcal{X}}$ be a real linear topological space, x_0 a vector of $\breve{\mathcal{X}}$ and $L(x)$ a real continuous functional on $\breve{\mathcal{X}}$ such that $L(x + y) \leq L(x) + L(y)$ and $L(\mathscr{b}x) = \mathscr{b}L(x)$ for $\mathscr{b} \geq 0$. Then there exist a continuous linear functional F on $\breve{\mathcal{X}}$ such that $F(x_0) = L(x_0)$ and $-L(-x) \leq F(x) \leq L(x)$ on $\breve{\mathcal{X}}$.*

The next theorem which is based on these theorems is introduced:

Theorem 4.4: *$\breve{\mathcal{X}}$ be a real linear topological space, x_0 a vector of $\breve{\mathcal{X}}$ and $L(x)$ a real linear continuous functional on $\breve{\mathcal{X}}$ such that $L(x + y) \leq L(x) + L(y)$ and $L(\mathscr{b}x) = \mathscr{b}L(x)$ for $\mathscr{b} \geq 0$. Then there exist a continuous linear affine functional F on $\breve{\mathcal{X}}$ such that $F(x_0) = L(x_0)$ and $F(x) = L(x) - c$ on $\breve{\mathcal{X}}$ where $c \in \mathbb{R}$.*

Proof from Theorem 3.3 states that $-L(-x) \leq F(x) \leq L(x)$. $F(x) \leq L(x)$, it implies that $F(x) = L(x) + q$,
where $q \in \breve{\mathcal{X}}$. If we use $q = -c$ and $c \in \breve{\mathcal{X}}$. thus,

$$F(x) = L(x) - c \tag{4.12}$$

Definition 4.3 and Equation (4.12) imply that for real linear continuous functional L there exists a continuous linear affine functional F on $\breve{\mathcal{X}}$ in the form

$$F(x) = L(x) - c \ \blacksquare.$$

Now based on previous theorems, the next theorem is introduced which is an important tool in defining abstract convex functionals.

Theorem 4.5: *Let $\mathcal{L}$ be the set of all linear functionals defined on $\mathbb{R}^n$ and $\mathcal{H}$ be the set of all affine functionals defined on $\mathbb{R}^n$. A functional H is affine if and only if $H(x) = L(x) - c$, where $L \in \mathcal{L}$ and $c \in \mathbb{R}$.*

Proof: While H is affine functional then From Definition 4.3, Equation (4.11) and $q = -c$ $(c \in \mathbb{R})$: $H(x) = L(x) + q \Rightarrow H(x) = L(x) - c$. Conversely, while $H(x) = L(x) - c$. Let $q = -c$ $(c \in \mathbb{R})$ then $H(x) = L(x) + q$. Based on Definition 4.17 H is affine functional.

The next theorem is stated.

Theorem 4.6: *A function $F: \mathbb{R}^n \to \mathbb{R}_{+\infty}$ is $\mathcal{H}$-convex if and only if F is a lower semicontinuous convex function.*

Proof: Equation (4.12) implies that left to right condition of the theorem is correct and Theorem 4.1 induce that right to left condition of the theorem is correct. On the other hand, it follows directly from Definition 4.2 that F is $\mathcal{H}$-convex if and only if $F(x) = \sup\{H(x): H \in \mathcal{H}, H \leq F, x \in \mathcal{X}\}$, where $H \leq F \Leftrightarrow (H(x) \leq F(x)$ for all $x \in \mathcal{X})$ ■.

Remark 4.1: Then a functional $P: \mathbb{R}^n \to \mathbb{R}_{+\infty}$ is $\mathcal{L}$-convex if and only if P is lower semicontinuous sublinear functional.

Then, the following definition is defined.

Definition 4.4: Let $F: \mathcal{X} \to \bar{\mathcal{V}}$. The set

$$\text{supp}\,(F, H) = \{H \in \mathcal{H}, H \leq F\} \tag{4.13}$$

of all $\mathcal{H}$-minorants of F is called the support set of the functional F with respect to the set of elementary functional H. Sometimes, we shall use the term lower support set instead of support set.

If L is the set of all linear functionals and P is a lower semicontinuous sublinear functional such that $P(0) < +\infty$. Then $P(0) = 0$ and

$$\text{supp}\,(P, L) = \{L \in \mathcal{L} : (\forall x \in \mathbb{R}^n) L(x) \leq P(x)\}$$
$$= \{L \in \mathcal{L} : (\forall x \in \mathbb{R}^n)\, L(x) - L(0) \leq P(x) - P(0)\} = \partial P(0),$$

where $P(0)$ is the subdifferential of the convex functional P at the origin. Note the following definition which is useful for proof of following lemma:

Definition 4.5: A set $\mathcal{U} \subset \mathcal{H}$ is called abstract convex with respect to $\mathcal{X}$ (or $(\mathcal{H}, \mathcal{X})$-convex) if there exists a functional $F: \mathcal{X} \to \overline{\mathbb{R}}$ such that $\mathcal{U} = supp(F, \mathcal{H})$.

The following lemma which is useful for considering the subdifferentiability of AC functionals is stated in next part.

Lemma 4.1: (Separation property) Let $\mathcal{H}$ be a set of elementary functionals defined on a set $\mathcal{X}$. Let $\mathcal{U}$ be a proper subset of $\mathcal{H}$, that is $\mathcal{U} \neq \emptyset$, $\mathcal{U} \neq \mathcal{H}$. Then the set $\mathcal{U}$ is $(\mathcal{H}, \mathcal{X})$-convex if and only if for each $H \notin \mathcal{U}$ there exists $x \in \mathcal{X}$ such that

$$H(x) > sup_{H' \in \mathcal{U}} H'(x) \tag{4.14}$$

Proof: similar to proof of lemma for functions by Rubinov (2000) let $\mathcal{U}$ be a proper $(\mathcal{H}, \mathcal{X})$-convex set. Let $P_{\mathcal{U}}$ is the upper envelope of the set $\mathcal{U}$ and

$$P_{\mathcal{U}}(x) = \sup\{H'(x) \colon H' \in \mathcal{U}\}.$$

It follows from the definition of abstract convex sets that $\mathcal{U} = \{H \in \mathcal{H} : H(x) \leq P_{\mathcal{U}}(x)\ for\ all\ x \in \mathcal{X}\}$. Thus if $H \notin \mathcal{U}$ there exists $x \in \mathcal{X}$ such that $H(x) > P_{\mathcal{U}}(x)$. Assume now that for each $H \notin \mathcal{U}$ one can fined $x \in \mathcal{X}$ such that (5.14) hold. Consider the set $\operatorname{supp}(P_{\mathcal{U}}, \mathcal{H})$. Clearly $\mathcal{U} \subset \operatorname{supp}(P_{\mathcal{U}}, \mathcal{H})$. If there exists $H \in \operatorname{supp}(P_{\mathcal{U}}, \mathcal{H})$ such that $H \notin \mathcal{U}$ then applying (4.14) and x' can be fined such that $H(x') > P_{\mathcal{U}}(x')$, which is impossible since $H(x) \leq P_{\mathcal{U}}(x)$ for all $x \in \mathcal{X}$ ∎.

4.9. Subdifferentiability of the Abstract Convex Functionals Defined on the Euclidean Spaces

As mentioned earlier there are two different approaches to subdifferentiability of convex functionals-local and global. The global definition can be naturally extended to abstract convex setting.

This section puts forward and extends appropriate definitions and propositions for functionals that Rubinov (2000) has for functions. Not that this section shall investigate only simple (finite elementary) functionals. Now the definition of the abstract subdifferential for abstract convex functionals is given.

Definition 5.6: Let $\mathcal{L}$ be a set of simple functionals defined on a set $\mathcal{X}$. A functional $L \in \mathcal{L}$ is called an abstract subgradient (or $\mathcal{L}$-subgradient) of a proper functional $F: \mathcal{X} \to \mathbb{R}_{+\infty}$ at a vector y if $F(x) \geq L(x) - (L(y) - F(y))$ for all $x \in \mathcal{X}$. The set $\partial_{\mathcal{L}} F(y)$ of all abstract subgradient of F at y is referred to as the abstract subdifferential (or $\mathcal{L}$-subdifferential) of the functional F at the vector y.

While the subdifferential $\partial_{\mathcal{L}} F(y)$ is not empty, it implies that

$$y \in dom\, F = \{x \in \mathcal{X}: F(x) < +\infty\}$$

This assertion follows directly from Definition 4.6.

Let $\mathcal{H}_{\mathcal{L}}$ be the closure of the set $\mathcal{L}$ under vertical shifts, that is

$$\mathcal{H}_{\mathcal{L}} = \{H': H'(x) = H(x) - c, H \in \mathcal{H}\}, c \in \mathbb{R}\}.$$

Clearly, $L \in \partial_{\mathcal{L}} F(y)$ if and only if $H_L \in supp(F, \mathcal{H}_{\mathcal{L}})$, where the functional

$$H_L(x) = L(x) - (L(y) - F(y)) \tag{4.15}$$

Is an element of $\mathcal{H}_{\mathcal{L}}$. The following assertion holds:

Proposition 4.2: Let $F: \mathcal{X} \to \mathbb{R}_{+\infty}$ be a proper functional. Then the subdifferential $\partial_{\mathcal{L}} F(y)$ is nonempty if and only if

$$F(y) = \max\{H(y): H \leq F, H \in \mathcal{H}_{\mathcal{L}}\}. \tag{4.16}$$

Proof: Similar to proof for function by Rubinov (2000), let $L \in \partial_{\mathcal{L}} F(y)$ and let H_L be a functional defined by (4.15). Then $H_L \in \mathcal{H}_{\mathcal{L}}$, $H_L \leq F$ and $H_L(y) = F(y)$. Hence (4.16) holds. Assume now that there exists functional $H \in supp(F, \mathcal{H}_{\mathcal{L}})$ such that $F(y) = H(y)$. Let $H(x) = L(x) - c$ with $L \in \mathcal{L}$ and $c \in \mathbb{R}$. Then $c = L(y) - F(y)$, hence the inequality $H(x) - (L(y) - F(y)) \leq F(x)$ holds ■.

Corollary 4.1: If a set $\mathcal{L}$ is closed under vertical shifts, that is $\mathcal{L} = \mathcal{H}_{\mathcal{L}}$ then

$$\partial_{\mathcal{L}} F(y) = \{L \in supp(F, \mathcal{L}): L(y) = F(y)\}. \tag{4.17}$$

Let $\mathcal{L}$ be the set of linear functionals defined on a vector space $\mathcal{X}$ then the set $\mathcal{H}_{\mathcal{L}}$ consists of affine functionals. Considering this observation the initial set $\mathcal{L}$ as a set of abstract linear functionals shall be investigated. Vertical shifts of functionals $L \in \mathcal{L}$, that is, functionals of the form

$$H(x) = L(x) - c \text{ for all } x \in \mathcal{X} \tag{4.18}$$

with $L \in \mathcal{L}$ and $c \in \mathbb{R}$ are called $\mathcal{L}$-affine functionals or abstract affine functionals with respect to $\mathcal{L}$. Therefore, the set $\mathcal{H}_{\mathcal{L}}$ is referred to as the set of $\mathcal{L}$-affine functionals. By property of (4.18) H and the pair $(L, c) \in \mathcal{L} \times \mathbb{R}$ is identified. It is possible that there are different pairs (L_1, c_1) and (L_2, c_2) such that $H(x) = L_1(x) - c_1 = L_2(x) - c_2$. For instance, if $\mathcal{L}$ is already closed under vertical shifts and $H(x) = L(x) - c$ then also is $H(x) = L'(x) - c'$ with $L' = L + d \in \mathcal{L}$, $c' = c + d \in \mathbb{R}$, where $1(x) = 1$ for all $x \in \mathcal{X}$ and $d \in \mathbb{R}$. Assume that the following condition (Σ) holds:

$$(\Sigma)\ L - c1 \notin \mathcal{L} \text{ for all } L \in \mathcal{L} \text{ and } c \neq 0 \tag{4.19}$$

It follows directly from (Σ) it is possible to represent an $\mathcal{L}$-affine functional H in the form (4.18) in a unique way. In such a case the set $\mathcal{H}_{\mathcal{L}}$ of all -affine functionals can be identified with $\mathcal{L} \times \mathbb{R}$. Let $\mathcal{L}$ be a set of abstract linear functionals defined on $\mathcal{X}$ and let $F: \mathcal{X} \to \mathbb{R}_{+\infty}$ be a proper functional. Assume that the subdifferential $\partial_{\mathcal{L}} F(x) \neq \emptyset$ for all $x \in \mathcal{X}$. Then, Proposition 4.1 induces that the functional F is abstract convex with respect to the set $\mathcal{H}_{\mathcal{L}}$ of abstract affine functionals and the supremum $\sup\{H(x): H \in supp(F, \mathcal{H}_{\mathcal{L}})\}$ is attained for each $x \in \mathcal{X}$. This observation establishes that the $\mathcal{L}$- subdifferential is an appropriate tool in the study of $\mathcal{H}_{\mathcal{L}}$-convex functionals.

Definition 4.7: A subset $\mathcal{X}$ of a vector space is called a cone set if $\mathscr{b}x \in \mathcal{X}$ for each $x \in \mathcal{X}$ and $\mathscr{b} > 0$.

The following proposition informs a very simple explanation of the $\mathcal{L}$- subdifferential for a set $\mathcal{L}$ consisting of positively homogeneous functionals.

Proposition 4.3: Let $\mathcal{L}$ be the set of abstract linear functionals defined on a conic set $\mathcal{X}$ and let each $L \in \mathcal{L}$ be positively homogeneous of degree one ($L(\mathscr{b}x) = \mathscr{b}L(x)\ for\ all\ \mathscr{b} > 0$). Then for an $\mathcal{L}$-convex functional P and for $\mathscr{y} \in \mathcal{X}$ we have:

$$\partial_{\mathcal{L}} P(\mathscr{y}) = \{L \in supp(P, \mathcal{L}): L(\mathscr{y}) = P(\mathscr{y})\} \tag{4.20}$$

Proof: Similar to proof on functions by Rubinov (2000), let $L \in \partial_{\mathcal{L}} P(\mathscr{y})$, that is, $P(x) \geq L(x) - (L(\mathscr{y}) - P(\mathscr{y}))$ for every $x \in \mathcal{X}$. Let $x \in \mathcal{X}$. Investigate vectors $\mathscr{b}x$ with $\mathscr{b} > 0$. Then

$$\mathscr{b}P(x) = P(\mathscr{b}x) \geq \mathscr{b}L(x) - (L(\mathscr{y}) - P(\mathscr{y})) \tag{4.21}$$

so,

$$\mathscr{b}P(x) = L(x) - \frac{L(\mathscr{y}) - P(\mathscr{y})}{\mathscr{b}}.$$

Passing to the limit as $\mathscr{b} \to +\infty$, then $P(x) \geq L(x)$ is gotten, that is, $L \in supp(P, \mathcal{L})$. Specially, $L(\mathscr{y}) \leq P(\mathscr{y})$ is stated. On the other hand, passing to the limit in (4.21) as $\mathscr{b} \to +\infty$, $L(\mathscr{y}) \geq P(\mathscr{y})$ is gotten. Therefore, $L(\mathscr{y}) = P(\mathscr{y})$ and $\partial_{\mathcal{L}} P(\mathscr{y})$ is contained in the set on the right-hand side in (4.20). Assume now that $P(x) \geq L(x)$ for all $x \in \mathcal{X}$ and $P(\mathscr{y}) - L(\mathscr{y}) = 0$. Then

$$P(x) \geq L(x) - (L(\mathscr{y}) - P(\mathscr{y})) \text{ for all } x \in \mathcal{X} \ \blacksquare.$$

Remark 4.2: A similar result is valid if $\mathcal{L}$ is closed under vertical shifts (see Corollary 4.1).

4.10. Introduction of Convex-Along-Rays Functionals on the Euclidean Spaces Based on Seyedi-Rohanin Model (SRM)

Based on Theorem 3.5 and SRM this section is going to propose CAR functional. This section shall put forward some theorems and propositions as stated by Rubinov (2000) which shall be extended to functionals and the proofs of them are one-to-one correspondence to their proof in the form of function.

First of all, the extension of convex-along-rays function definition in functionals is proposed.

Definition 4.8: Let $Q \subset \mathbb{R}^n$ be a cone. A functional $F: Q \to \mathbb{R}_{+\infty}$ is called convex-along-rays (CAR) if, for each $x \in Q$, the functional

$$F_x(t) = F(tx) \qquad t \in [0, 1) \tag{4.22}$$

is convex.

The closed ray $\mathcal{R}_x$ is denoted by the set: $\{tx: 0 \leq t < +\infty\}$. This definition is extended in the following definition.

Definition 4.9: A functional $F: Q \to \mathbb{R}_{+\infty}$ is called convex-along-rays (CAR) if its restriction to each ray $\mathcal{R}_x$ with $x \in Q$ is a convex functional.

In this part the increasing convex-along-rays (ICAR) functionals defined on $Q = \mathbb{R}_+^n$ or $Q = \mathbb{R}_{++}^n$ is investigated. Therefore, the following definition is stated.

Definition 5.10: Let Q be either $\mathbb{R}_+^n$ or $\mathbb{R}_{++}^n$. A functional $F: Q \to \mathbb{R}_{+\infty}$ is an ICAR functional if the following conditions hold:

- F is increasing: $x \geq y$ implies $F(x) \geq F(y)$;
- for each $x \in Q$, the functional F_x defined by (4.22) is convex.

By continuing this way, some examples of ICAR functionals are:

(a) An increasing convex functionals defined on $\mathbb{R}_+^n$ is ICAR.
(b) An increasing positively homogeneous functional of degree $m \geq 1$ defined on $\mathbb{R}_+^n$ is ICAR. Especially,
(c) An IPH functional defined on $\mathbb{R}_+^n$ is ICAR.

The following propositions are useful for showing the properties of ICAR functionals.

Proposition 4.4: The set $\mathcal{F}$ of all ICAR functionals defined on either $\mathbb{R}_+^n$ or $\mathbb{R}_{++}^n$ is a convex cone.

Proof: Directly from the definition it is easy to show that if F_1, F_2 is ICAR functionals and $b_1, b_2 > 0$ then the functional $b_1 F_1 + b_2 F_2$ is ICAR ■.

Proposition 4.5: Let F be an increasing functional on $\mathbb{R}^n_+$. Then the level sets $\{x \in \mathbb{R}^n_+ : F(x) \leq c\}$ are normal and the level sets $\{x \in \mathbb{R}^n_+ : F(x) \geq c\}$ are conormal. Specially, the set $dom\, F = \{x \in \mathbb{R}^n_+ : F(x) < +\infty\}$ is normal and the set $\{x \in \mathbb{R}^n_+ : F(x) = +\infty\}$ is conormal.

Proof: The proof follows directly from the definition 3.10 ■.

Proposition 4.6: Any $\mathcal{H}_\mathcal{L}$-convex functional $F: \mathbb{R}^n_+ \to \mathbb{R}_{+\infty}$ is lower semicontinuous and ICAR.

Proof: Let F be an $\mathcal{H}_\mathcal{L}$-convex functional, that is, $F(x) = \sup\,\{H(x): H \in supp\,(F, \mathcal{H}_\mathcal{L})\}$. Then for each $x \in \mathbb{R}^n_+$: $F_x(t) = \sup\{H_x(t): H \in supp(F, \mathcal{H}_\mathcal{L})\}$, where $F_x(t)$ and $H_x(t)$ are defined as in (4.22). Note that the functional H_x is affine for each $H \in \mathcal{H}_\mathcal{L}$ and $x \in \mathbb{R}^n_+$, hence the functional F_x is convex. Thus F is convex-along-rays. Since each functional $H \in \mathcal{H}_\mathcal{L}$ is increasing, it follows that F is increasing as well. Since each $H \in \mathcal{H}_\mathcal{L}$ is a continuous functional, it implies that, which is the upper envelope of the set supp $(F, \mathcal{H}_\mathcal{L})$ of continuous functionals, is lower semicontinuous ■.

The next assertion informs that the class of ICAR functionals is very large. But before that performance of one lemma is useful.

Lemma 4.2: Let $\mathcal{H}$ be a set of continuous functionals defined on a metric space $\breve{\mathcal{Z}}$ with the following properties:

(i) $\mathcal{H}$ is a conic set;

(ii) $H + \mathscr{b}\mathbf{1} \in \mathcal{H}$ for each $H \in \mathcal{H}$ and $\mathscr{b} < 0$ (here $1(x) = 1$ for all $x \in \breve{\mathcal{Z}}$);

(iii) For each $z \in \breve{\mathcal{Z}}$ there exists a functional $H \in \mathcal{H}$ such that $H(z) = 0$, $H(x) < 0$ for $x \neq z$, $H + \sigma\mathbf{1} \in \mathcal{H}$ for all small enough $\sigma > 0$.

Then a functional F defined on $\check{Z}$ is H-convex if and only if F is lower semicontinuous [9].

Thus based on this lemma, the following proposition shall be stated.

Proposition 4.7: Let F be a lower semicontinuous functional defined on the unit simplex $\mathcal{S}^* = \{x \in \mathbb{R}^n_+ : \sum_i x_i = 1\}$. Then there exists an ICAR extension of F, that is, an ICAR functional $\bar{F}: \mathbb{R}^n_+ \to \mathbb{R}_{+\infty}$ such that $\bar{F}(x) = F(x)$ for all $x \in \mathcal{S}^*$.

Proof: Similar to proof of functions let $\mathcal{H}_{\mathcal{S}^*}$ be the set of all functionals $H_{\mathcal{S}^*}$ defined on the simplex $\mathcal{S}^*$ by $H_{\mathcal{S}^*}(x) = \langle \ell, x \rangle - c$ with $\ell \in \mathbb{R}^n_+$, $c \in \mathbb{R}$ where $\langle \ell, x \rangle = \min_{i \in \mathcal{I}_+(l)} \ell_i x_i$ and $\mathcal{I}_+(\ell) = \{i : \ell_i > 0\}$. Obviously, conditions (i) and (ii) from Lemma 4.2 hold for the set $\mathcal{H}_{\mathcal{S}^*}$. In addition, the condition (iii) can be checked. Let $z \in \mathcal{S}^*$. Consider the vector $\ell = 1/z$, where

$$\left(\frac{1}{z}\right)_i = \begin{cases} \dfrac{1}{z_i} \ if \ z_i > 0; \\ 0 \ if \ z_i = 0. \end{cases}$$

It is obvious that $\langle \ell, z \rangle = 1$. Since

$$\sum_{i=1}^{n} x_i = \sum_{i=1}^{n} z_i$$

for $x \in \mathcal{S}^*$, it induces that for $x \neq z$ there exists an index j such that $x_j < z_j$. Obviously, $j \in \mathcal{I}_+(z) = \mathcal{I}_+(\ell)$. Hence,

$$\langle \ell, x \rangle = \min_{i \in \mathcal{I}_+(z)} \frac{x_i}{z_i} < 1.$$

Consider the functional H defined on $\mathcal{S}^*$ by $H(x) = \langle \ell, x \rangle - 1$. Then

$$H(z) = 0 \text{ and } H(x) < 0 \text{ for all } x \neq z$$

Also,

$$H + \sigma.\mathbf{1} \in \mathcal{H}_{\mathcal{S}^*} \text{ for all } \sigma.$$

Thus the condition (iii) of Lemma 4.2 holds. Let $F: \mathcal{S}^* \to \mathbb{R}_{+\infty}$ be a lower semicontinuous functional. It implies that from Lemma 4.2 that there exists a set $\mathcal{U} \subset \mathcal{H}_{\mathcal{S}^*}$ such that $F(x) = sup_{H \in \mathcal{U}} H(x)$ for all $x \in \mathcal{S}^*$. Investigate now the functional $\tilde{F}$ defined on $\mathbb{R}^n_+$ by

$$\tilde{F}(x) = \sup\{H(x): H \in \mathcal{U}\}.$$

It follows from Proposition 4.6 that $\tilde{F}$ is an ICAR functional. Also,

$$\tilde{F}(x) = F(x) \text{ for all } x \in \mathcal{S}^* \ \blacksquare.$$

Then the following propositions stated the relationship between ICAR functional and $\mathcal{H}_{\mathcal{L}}$-convex functional based on lower semicontinuous functionals.

Proposition 4.8: Let $F: \mathbb{R}^n_+ \to \mathbb{R}_{+\infty}$ be an ICAR functional such that for $x \in \mathbb{R}^n_+$ the functional F_x defined by (4.22) is lower semicontinuous on $\mathbb{R}_+$. Then the functional F is abstract convex with respect to the set $\mathcal{H}_{\mathcal{L}}$ of all functionals H defined by

$$H(x) = \langle \ell, x \rangle - c \text{ and } \ell, x \in \mathbb{R}^n_+. \tag{4.23}$$

Proof: Proof of the theorem in functionals is similar to functions by Rubinov (2000) completely ∎.

The following theorem is investigated based on Propositions 4.6, 4.7 and 4.8.

Theorem 4.7: *Let $\mathcal{H}_{\mathcal{L}}$ be the class of all functionals H by (5.23). A functional $F: \mathbb{R}_+^n \to \mathbb{R}_{+\infty}$ is $\mathcal{H}_{\mathcal{L}}$-convex if and only if F is lower semicontinuous and ICAR.*

Proof: Let F be a lower semicontinuous ICAR functionals. Then the functional F_{y} is lower semicontinuous for each y. Thus Proposition 4.8 demonstrates that F is $\mathcal{H}_{\mathcal{L}}$-convex. The reverse assertion follows from Proposition 4.6 ∎.

The following proposition will be useful for subdifferential of ICAR functionals.

Proposition 4.9: Let F ba an ICAR functional and $y \in \mathbb{R}_+^n \backslash \{o\}$ be a vector, such that $(1+\varepsilon)y \in domF$ for some $\varepsilon > 0$. Then there exists $\ell \in \mathbb{R}_+^n$ with the property

$$\langle \ell, x \rangle - \langle \ell, y \rangle \leq F(x) - F(y) \text{ for all } x \in \mathbb{R}_+^n.$$

In particular the vector $\ell = u/y$ with $u \in \partial F_{y}(1)$ enjoys this property.

Proof: Proof of the theorem in functionals is similar to functions by Rubinov (2000) completely.

4.11. Subdifferentiability of Increasing Convex-Along-Rays Functionals

Abstract convexity with respect to the set $\mathcal{H}_{\mathcal{L}}$ of all $\mathcal{L}$-affine functionals is investigated, where $\mathcal{L}$ is the set of all min-type functionals defined on $\mathbb{R}^n_+$. According Theorem 4.7 and Proposition 4.9 it is easy to prove the following proposition.

Proposition 4.10: A functional $F: \mathbb{R}^n_+ \to \mathbb{R}_{+\infty}$ is abstract convex with respect to $\mathcal{H}_{\mathcal{L}}$ if and only if F is lower semicontinuous and ICAR.

Proof: It follows from Theorem 4.7 directly ■.

By continuing this way the $\mathcal{L}$-subdifferential $\partial F_{\mathcal{L}}(x)$ of an ICAR functionals F at a vector $x \in dom\ F$ shall be studied. Recall that

$$\partial F_{\mathcal{L}}(x) = \{\ell \in \mathcal{L}: \langle \ell, y\rangle - \langle \ell, x\rangle \leq F(y) - F(x)\}.$$

Then,

Theorem 4.8: *Let F be an ICAR functional and* $x \in \mathbb{R}^n_+ \backslash \{o\}$ *such that* $(1+\varepsilon)x \in domF$ for some $\varepsilon > 0$. *Then the subdifferential* $\partial F_{\mathcal{L}}(x)$ *is not empty and*

$$\{u/x : u \in \partial F_x(1)\} \subset \partial F_{\mathcal{L}}(x), \tag{4.24}$$

where $F_x(t) = F(tx)$.

Proof: the result follows directly from Proposition 4.9 ■.

Since F is increasing it follows that $o \in \partial F_{\mathcal{L}}(o)$ therefore, $\partial F_{\mathcal{L}}(o)$ is also nonempty.

Let F be a finite ICAR functional. For $x \in \mathbb{R}^n_+$ the following vector is investigated

$$F^{\nabla}(x) = \frac{F'(x,x)}{x}. \tag{4.25}$$

It follows immediately from Theorem 4.8 that $F^{\nabla}(x) \in \partial F_{\mathcal{L}}(x)$ for all $x \in \mathbb{R}^n_+$.

REFERENCES

[1] Bylinski, C. (1989). "Some basic properties of sets", *Journal of Formalized Mathematics*, 1(198), 9.

[2] Hryniewiecki, K. (1990). "Basic properties of real numbers", *Formalized Mathematics*, 1(1), 35-40.

[3] Kotowicz, J. & Sakai, Y. (1992). "Properties of partial functions from a domain to the set of real numbers", *Formalized Mathematics*, 3(2), 279-288.

[4] Nowak, B. & Trybulec, A. (1993). "Hahn-Banach theorem", *Journal of Formalized Mathematics*, 5(199), 3.

[5] Hamel, A. (2004). "From real to set-valued coherent risk measures", *Reports of the Institute of Optimization and Stochastics*, Martin-Luther-University Halle-Wittenberg, 19, 10-22.

[6] Anger, B. (1977). "Representation of capacities", *Mathematische Annalen*, 229(3), 245-258.

[7] Kurdila, A. J. & Zabarankin, M. (2005). "*Convex Functional Analysis, Systems and Control: Foundations and Applications*," ed: Birkhäuser, Basel.

[8] Koliha, J. J. & Leung, A. (1975). "Ergodic families of affine operators", *Mathematische Annalen*, 216(3), 273-284.

[9] Rubinov, A. M. (2000). *Abstract convexity and global optimization* vol. 44, Springer.

Chapter 5

CAPABILITY OF FUNCTION OPTIMIZATION ALGORITHMS FOR SOLVING OPTIMAL CONTROL PROBLEMS WITH RESPECT TO THE INHERITANCE AND GENERALIZABILITY PROPERTIES

ABSTRACT

An optimization algorithm as a numerical method is used to find the optimum values of functions. This study investigates the existence of the Inheritance and Generalizability properties on optimization algorithms. In particular, based on these properties, this study shows that optimization algorithms which work for functions can be extended to solve functionals.

Keywords: subset principle, axiom of induction, function, functional, optimization algorithm

5.1. Introduction

Calculus of variations is a subdivision of mathematics which includes a type of generalization of calculus. It seeks to find the path, curve, surface, etc., for which a given function has a stationary value (which, in physical problems, is usually a minimum or maximum).

In mathematics, this involves finding stationary values of integrals of the form.

$$I = \int_a^d F(y, \dot{y}, x)dx,$$

which is called a functional.

I has an extremum if the Euler-Lngrange differential is satisfied, i.e., if $\frac{\partial F}{\partial y} - \frac{d}{dx}\left(\frac{\partial F}{\partial \dot{y}}\right) = 0$.

Optimal control is the process of finding control and state histories for a dynamic system over a period of time to minimise a performance index. Optimal control is used in many fields, for example:

- To determine efficient manoeuvre of aircraft, spacecraft, and robots.
- To design feedback controllers which are optimal in some sense [1].

In most branches of science, finding optimal value of the functionals is becoming an increasingly important field of study. For this reason, many scientists proposed several appropriate optimization methods to find local or global optimum of the functionals. This research is going to demonstrate properties which exist the optimization algorithms in this field. As mentioned before, the properties of the functional (which are inherited from function naturally) caused the optimization algorithms that are used for functions can also be extended for use in functionals. Definitions, theorems and examples shall be put forward below, in investigating these properties.

5.2. PRELIMINARIES FROM THE SET THEORY

Let set $\mathcal{D}$ is the set of all optimization algorithms which are used for objective functions and set $\mathcal{Q}$ as the set of all optimization algorithms which are used for objective functionals (see Figure 5.1). According to the discussion in Chapter 3, it is clear that the set $\mathcal{Q}$ is a subset of set $\mathcal{D}$. Therefore, a functional is a type of function and the optimization algorithms that are used for functions can also be used for functionals. According to Theorem 3.4, set $\mathcal{Q}$ inherits the properties and laws of set $\mathcal{D}$. It is very easy to verify that these sets are countably finite.

All of the algorithms that are used for functions in set $\mathcal{D}$ can be indexed as $\{d_i| i=1,...,n;\ n \in \mathbb{N}\}$and all of the algorithms used for functionals in set $\mathcal{Q}$ can be indexed as $\mathcal{Q}=\{q_j|j = 1, \dots, m;\ m \in \mathbb{N}$ $\}$where $\mathcal{Q} \subseteq \mathcal{D}$. Similarly, two bijective maps (π_1, π_2) can be described from the set of natural numbers to both sets by denoting that they are countably finite. Hence a map (π) from set $\mathcal{D}$ to $\mathcal{Q}$ is bijective (see Figure 5.2).

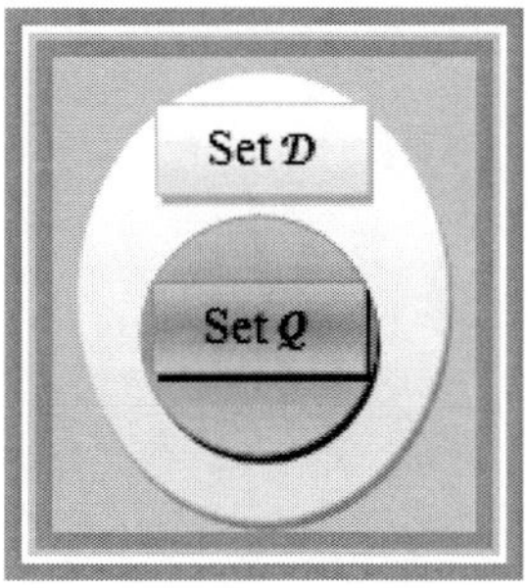

Figure 5.1. The set of all optimization algorithms which are used for objective functionals is the subset of the set all optimization algorithms which are used for objective functions, [2].

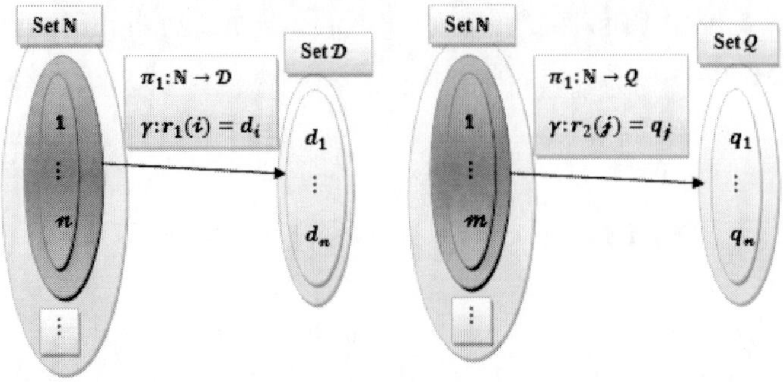

Figure 5.2. Performance of countability of set $\mathcal{D}$ (all function optimization algorithms) and set $\mathcal{Q}$ (all functional optimization algorithms) [2].

From this observation, the following theorem shall be investigated.

Theorem 5.1: *Let the set* $\mathcal{D}$ *be the set of all optimization algorithms which are used for functions and* $\mathcal{Q}$ *be the set of all optimization algorithms which are used for functionals. If* (π_1, π_2) *are bijective maps from the natural number to both sets, then* π *as a map from set* $\mathcal{D}$ *to* $\mathcal{Q}$ *is also bijective [2].*

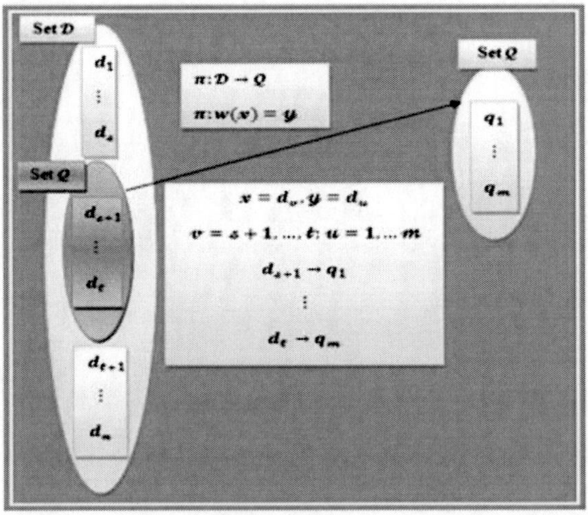

Figure 5.3. The existence of bijective map from the set of all functional optimization algorithms which is the subset of the set of all function optimization algorithms onto itself, [2].

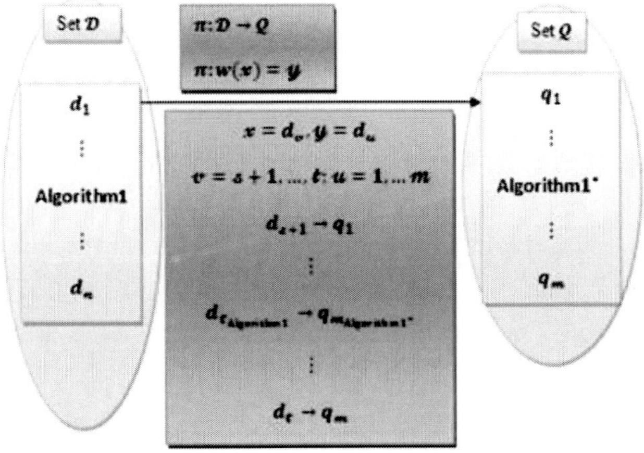

Figure 5.4. An optimization algorithm used for functions can also be used for functional, [2].

Proof: the proof completely is similar to the proof of Theorem 3.3 (see Figures 5.3 and 5.4)■.

5.3. The Inheritance and Generalizability Properties in the Structure of the Optimization Techniques

In this section, properties inherited from functions shall be discussed. The discussion shall begin with the following theorem on subset principle.

Theorem 5.2 (Subset Principle): *If* $\Xi^*(x)$ *is a condition on sets, then, for each set* $\mathcal{A}^*$, *there exists a unique set whose members are precisely those members* $\mathcal{Z}^*$ *of* $\mathcal{A}^*$ *for which* $\Xi^*(\mathcal{Z}^*)$ *holds.*

From Theorem 5.2, let Ξ^* (.) be the Laws and Properties of optimization algorithms which are used for functions.

According to Theorem 5.2, it is clear that subsets inherit their properties from their superset. Therefore, it implies that optimization algorithms that are used for functionals should observe laws and properties of optimization algorithms that are used for functions.

Remark 5.1 (Inheritance Property): According to Theorem 5.2, it is clear that subsets inherit their properties from their superset. It implies that any optimization algorithms which are used for functionals should observe all function optimization algorithms laws and properties as its inherited property.

Remark 5.2 (Generalizability Property): Theorems 5.2 and 3.3 and SRM permit the presentation of any type of function optimization algorithms to use for functionals and conversely.

This section investigates the existence of the properties in the optimization algorithms theoretically. Furthermore, the next part shall be demonstrating these properties numerically. The following section shows the capability of some major optimization algorithms which are applied for both functions and functionals. To exemplify this fact some numerical examples shall be mentioned. Next based on the properties and SRM, the cutting angle method shall be presented for use in functionals.

5.4. Some Numerical Examples of the Capability of the Function Optimization Algorithms to Use for Optimizing the Functionals

The previous section investigated the properties of functional optimization algorithms theoretically based on SRM. This section is going to demonstrate these properties in some well-known optimization algorithms numerically. Then each subsection here shall be demonstrating the application of the algorithms both in function and functional for accentuating the existence of these properties.

5.4.1. Simplex Method

The simplex algorithm, created by the American mathematician George Dantzig in 1947, is a popular algorithm for numerical solution of the linear programming problem. The journal Computing in Science and Engineering listed it as one of the top 10 algorithms of the century.

The following Example 5.1 is an example which was solved by the method. The method was used for functions successfully. The following Example 6.1 from Swarup and Bedi (1972) is a successful usage of the method on nonlinear optimization.

Example 5.1: Min

$$f(x) = -6x_1 + 2x_1^2 - 2x_1x_2 + 2x_2^2$$

subject to

$$x_1 + x_2 \leq 2$$

$$x_1, x_2 \geq 0.$$

The optimal solution to the problem was $x_1 = 3/2, , x_2 = \frac{1}{2}$ and the corresponding value of the objective function was $f(x) = -11/2$.

The method can also be extended for use in functionals. This is evident from examples are gathered in Fakharzadeh, et al. (2008), the method was used for solving optimal control problems. An excerpt of the examples is given in Example 6.2 below.

Example 5.2: Minimize

$$\int_0^1 u^2(t)\,dt$$

$$\dot{x}(t) = \frac{1}{2}x + u,$$

$$x(0) = 0.$$

This problem was solved by revised simplex method with subroutine DLPRS in IMSL library of Compaq Visual Fortran6. The optimal value of objective function was 0.145218.

5.4.2. Newton Method

Newton's method is a well-known algorithm for finding roots of equations in one or more dimensions. It can also be used to find local maxima and local minima of functions by noticing that if a real number $x^*$is a stationary vector of a function $f(x)$, then x^* is a root of the derivative $f(x)$, and therefore one can solve for x^* by applying Newton's method to $f(x)$. The Taylor expansion of $f(x)$,

$$f(x + \Delta x) = f(x) + f'(x)\Delta x + \frac{1}{2}f''(x)\Delta x^2,$$

attains its extremum when Δx solves the linear equation:

$$f'(x) + f''(x)\Delta x = 0.$$

Thus, provided that $f(x)$ is a twice-differentiable function and the initial guess x_0 is chosen close enough to x^*, the sequence (x_n) defined by

$$x_{n+1} = x_n - \frac{f'(x_{n+1})}{f''(x_n)}, n \geq 0$$

will converge towards x^*.

The above discussion shows that Newton Method can be used as an optimization algorithm for finding the minimizer or maximizer of functions. The following problem is an example solved by using Newton Optimization Method.

Example 5.3: Consider $f(x) = x^3 + 4x^2 - 10$, Newton Optimization Method was used with several initial solution. After a number of iterations the minimizer of $f(x)$ was obtained. The numerical results are listed in Table 5.1, with i being its iteration number.

Similarly, the following are examples showing the application of Newton method for minimization of functionals.

To solve the following problem, Chebyshev polynomials technique was used for discretizing the objective. Example 5.4 was carried out on VAX/8530 and CDC Cyber 170/750, by Kazemi and Miri (1993).

Example 5.4: The objective is to find the optimal control $\mathrm{u}(t)$ which minimizes the energy cost functional

$$J = \frac{1}{2}\int_0^1 (\mathrm{x}^2 + \mathrm{u}^2)dt \tag{5.1}$$

Table 5.1. Value of minimizer x^* of the $f(x)$ with initial point x_0 in iteration i, [2]

x_0	i	x^*	$f(x^*)$
0.5	4	3.9797e-013	10
1	4	2.4982e-009	10
2	5	4.4742e-012	10
– 0.3	4	1.1850e-014	10

subject to

$$\dot{x} = -x + u, x(0) = 1, \quad (5.2)$$

The optimum solution is $J = 0.19290922$, For $N = 6$, where p is the order of Chebyshev polynomials technique for solving (5.1)-(5.2) (see Table 5.2).

Example 5.5 was solved using subroutine E04NCF of the NAG library, by Alt and Malanowski (1995).

5.4.3. Genetic Algorithm

A genetic algorithm (GA) is a search technique used to compute the approximate solutions for optimization and search problems. Genetic algorithms are categorized as global search heuristics. Genetic algorithms are a particular class of evolutionary algorithms that use techniques inspired by evolutionary biologies such as inheritance, mutation, selection, and crossover (also called recombination).

The following examples shall show that the GA was used for functions and functionals successfully. The immediate example demonstrates the capability of GA for minimizing functions.

An example used in the literature for genetic algorithms testing by Shopova and Vaklieva-Bancheva (2006) is presented below as an illustration. All simulations were done on Pentium-IV machine.

Table 5.2. Performance index for different values of p. [3]

p	4	5	6
J	0.19289779	0.19290691	0.19290922

Example 5.5: Minimize

$$(y_1-1)^2+(y_2-2)^2+(y_3-1)^2-\log(y_4+1)+(x_1-1)^2 \\ +(x_2-2)^2+(x_3-3)^2$$

subject to

$$y_1+y_2+y_3+x_1+x_2+x_3\le 5$$

$${y_3}^2+{x_1}^2+{x_2}^2+{x_3}^2\le 5.5$$

$$y_1+x_1\le 1.2$$

$$y_2+x_2\le 1.8$$

$$y_3+x_3\le 2.5$$

$$y_4+x_1\le 1.2$$

$${y_2}^2+{x_2}^2\le 1.64$$

$${y_3}^2+{x_3}^2\le 4.25$$

$${y_2}^2+{x_3}^2\le 4.64$$

$$x_i\ge 0, i=1,\dots,3$$

$$y_i\in\{0,1\}, i=1,\dots,4$$

It is a MINLP minimization problem taken from Floudas, et al. (1989). It has four binary and three continuous variables with nine inequality constraints. The obtained optimum solution with Genetic

Algorithm was 4.971699587 by Shopova and Vaklieva-Bancheva (2006).

GA was used for minimizing functionals successfully. The following examples were minimized by GA and were discretized by Chebyshev polynomials technique.

Fard and Borzabadi (2007) used the Genetic Algorithm Toolbox of Matlab (Ver, 7.04) to solve the examples.

Example 5.6: Fard and Borzabadi (2007) considered the following optimal control problem

Minimize

$$I = \int_0^1 u^2(t)dt$$

subject to

$$\dot{x}(t) = x^2(t) + u(t)$$
$$x(0) = 0, x(1) = 0.5.$$

After solving with the Genetic Algorithm, Fard and Borzabadi (2007) obtained the following results:

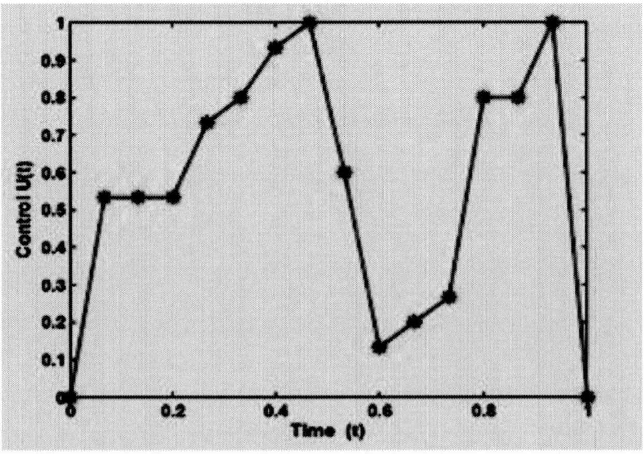

Figure 5.5. Control function for Example 5.6 [4].

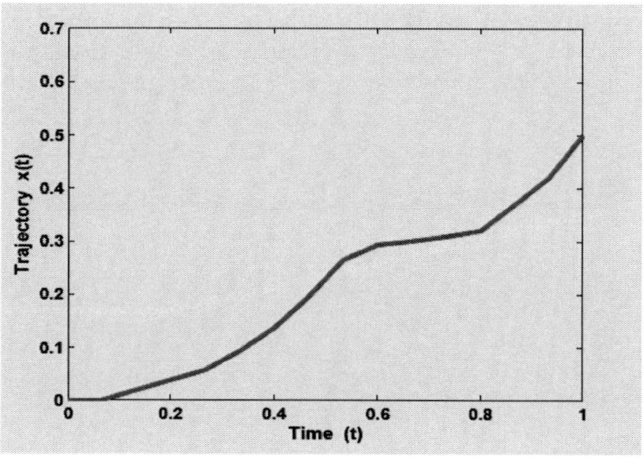

Figure 5.6. Trajectory function for Example 5.6 [4].

The final value $\mathrm{x}(1) = 0.500054$, the optimal value $I^* = 0.4447$ and the error function $e(1) = 5.43 \times 10 - 5$. The control and trajectory functions are shown in Figure 5.5 and 5.6, respectively.

Example 5.7: Fard and Borzabadi (2007) considered the following optimal control problem

Minimize

$$I = \int_0^1 \mathrm{u}^2(t)dt$$

subject to

$$\dot{\mathrm{x}}(t) = \frac{1}{2}\mathrm{x}^2(t)\sin(\mathrm{x}(t)) + u(t)$$

$$\mathrm{x}(0) = 0, \mathrm{x}(1) = 0.5.$$

After solving the problem with Genetic Algorithm, Fard and Borzabadi (2007) obtained the following results:

The final value x(1) = 0.500040, the optimal value I^* = 0.3526 and the error function $e(1)$ = 4.095 × 10 − 5. The control and trajectory functions are shown in Figure 5.7 and 5.8, respectively.

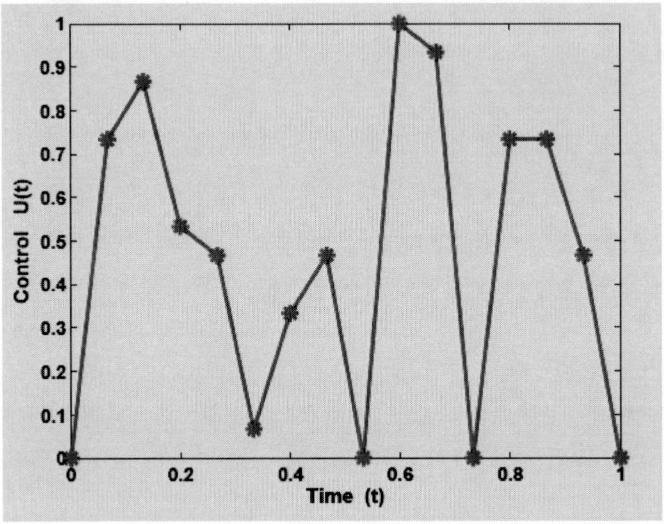

Figure 5.7. Control function for Example 5.7 [4].

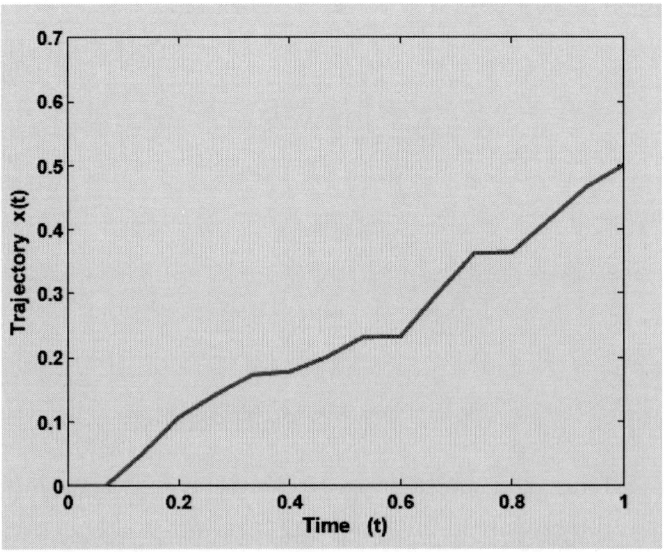

Figure 5.8. Trajectory function for Example 5.7 [4].

REFERENCES

[1] Becerra, V. M. (2008). "Optimal control", *Scholarpedia*, 3(1), 5354.

[2] Seyedi, S., Ahmad, R. & Aziz, M. I. A. (2009). *"Inheritance of Function Properties for Functionals"*.

[3] Kazemi, M. & Miri, M. (1993). "Numerical solution of optimal control problems", in *Southeastcon'93*, *Proceedings*, *IEEE*, 2 p.

[4] Fard, O. S. & Borzabadi, A. H. (2007). "Optimal control problem, quasi-assignment problem and genetic algorithm", in *Proceedings of World Academy of Science*, *Engineering and Technology*, 70-43.

Chapter 6

A Generalized Version of Cutting Angle Method for Solving Continuous Time Optimal Control Problems

Abstract

This study investigates an application of the cutting angle method to a class of problems referred to as the optimal control problems. The solution approach here is given with respect to a new technique called unit vector combinations. At the end, based on this approach, some numerical examples are driven.

Keywords: continuous time-optimal control problem, optimization, cutting angle method

6.1. Introduction

Optimal control is a study which deals with selecting the best set of decisions to achieve a particular objective [1]. The performance of physical processes depends on a great deal of these decisions. The static process and the dynamic process are two classifications of Physical processes. The dynamic process can be further segregated into two linear and the non-linear processes. The main focus of this study is on the non-linear dynamic process.

Following modeling a linear process through optimization of the non-linear dynamic process, it is solved iteratively by applying an appropriate dynamic optimization algorithm. Despite that there are some inaccuracies in the mathematical models employed, the solution of the linear process will converge to the solution of the original non-linear problem [2, 3].

Although countless algorithms are accessible in order to solve Optimal Control Problems (OCP), there are essentially local search algorithms. To solve the problem associated with local searches, most OCP is modeled as Linear Quadratic Regulator (LQR) problems. So it is hoped that this newfound solution can be considered as the true global solution to the original problem. However, when it is done, a great deal of information which is carried by the original problem might be lost in its translation into LQR models.

This paper shall introduce a dynamic global search algorithm called Cutting Angle Method (CAM). This method was introduced by Bagirov and Rubinov (2000) which is a generalization of the cutting plane method from convex minimization, for solving OCP. CAM has been successfully applied to functions, whereas OCP is functionals.

According to Bagirov and Robinov (2000), CAM can be possibly applied to a noticeably broad class of non-convex global optimization problems. Bagirov and Rubinov (2000) claimed that CAM used in combination with a local search algorithm to arrive at a truly global solution can help local search algorithms to converge to global optima. Basically, the approach begins with a solution of a local search algorithm. The cutting angle method is then applied and repeated until one arrives at a solution that is lower than the current solution of the local search algorithm. While the lower solution as the new initial solution is employed, the local search algorithm is again run on the problem. As soon as a new solution is obtained from this local search, the cutting angle method is once again used and the process would not stop unless the cutting angle method fails to find a lower valued solution after a predetermined number of iterations. The newest solution of the local search algorithm would be considered as the global optimum. This research mainly concerns on proposing a technique to employ CAM for solving OCP directly.

This chapter is going to develop an application of CAM for solving OCP consisting of two steps:

Step 1. Discretize the OCP, and
Step 2. Solve the approximated problem using the CAM via Unit Vectors Combinations Technique (UVCT).

Convergence of the solutions of the Discretized OCP problems to extremal via CAM is based on the convergence of the CAM to global solution by Andramonov et al. (1999).

6.2. An Application of the Cutting Angle Method to Solve Continuous Time OCP

The next procedure shall analyze an application of CAM to OCP. This application is generated from two approaches which were introduced by Bagirov and Rubinov (2000), discretization techniques and the Inheritance and Generalizability of function optimization algorithms for solving functional (In Particular, OCP).

Procedure 6.1

(I) (Discretization): The original problem in this research is in the form of:

(II)

$$\text{Minimize } J = \int_{t_0}^{t_1} E(\boldsymbol{x}, \boldsymbol{u}, t)dt \tag{6.1}$$

$$s.t\ \dot{\boldsymbol{x}} = g(\boldsymbol{x}, \boldsymbol{u}, t) \tag{6.2}$$

$$\boldsymbol{x}(t_0) = \boldsymbol{x}_0, \boldsymbol{x}(t_1) = \boldsymbol{x}_1,$$

where $E: \mathcal{A} \times \mathcal{B} \times [t_0, t_1] \to \mathbb{R}$ is a continuous function, g is a continuous nonlinear time varying function, $g: \mathcal{A} \times \mathcal{B} \times [t_0, t_1] \to \mathbb{R}^n, t \in [t_0, t_1] \subseteq \mathbb{R},$, $\boldsymbol{x}(t) \in \mathcal{A}$, $\boldsymbol{u}(t) \in \mathcal{B}$ and $\mathcal{A} \subseteq \mathbb{R}^n, \mathcal{B} \subseteq \mathbb{R}^m$ are compact subset and must be chosen so that the system reach from initial state $\boldsymbol{x}(t_0)$ to final state $\boldsymbol{x}(t_1)$.

Firstly, discretize the original problem (6.1) into the following equivalent problem and then partition the interval $[t_0, t_1]$ into n subsets and also $t \in [t_0, t_1]$.

$$\text{Minimize } J = \mathfrak{h} \sum_{i=1}^{n-1} E_i\,(\boldsymbol{x}, \boldsymbol{u}, t) \tag{6.3}$$

$$s.t\ \mathrm{x} = \mathfrak{h} * g(\boldsymbol{x}, \boldsymbol{u}, t) \tag{6.4}$$

$$\boldsymbol{x}(t_0) = \boldsymbol{x}_0,, \boldsymbol{x}(t_1) = \boldsymbol{x}_1,$$

where $\mathfrak{h}$ is a step size of the discretization, $\mathfrak{h} = t_1 - t_0/n$, $\boldsymbol{x} = \{\boldsymbol{x}_j\}, \boldsymbol{u} = \{\boldsymbol{u}_k\}, j = 1,2, \dots, m_1$ and $k = 1,2, \dots, m_2$.

(II) Secondly, apply CAM for solving the discretized OCP (6.3) via unit vectors Combinations technique (UVCT). This research

generalizes CAM to solve discretized OCP via the UVCT as follows,
In Step 0 of CAM initialize the state vector $\boldsymbol{x}_1$ by unit vectors

$$e^1 = (1,0,0,\dots,0,0)_{1\times n-1}, e^2 = (0,1,0,\dots,0,0)_{1\times n-1},\dots,$$
$$e^{n-2} = (0,0,0,\dots,1,0)_{1\times n-1}, e^{n-1} = (0,0,0,\dots,0,1)_{1\times n-1}.$$

Next, calculate the corresponding values of $\boldsymbol{x}_2, \boldsymbol{x}_3, \dots, \boldsymbol{x}_{m_1}$; $\boldsymbol{u}_1, \boldsymbol{u}_2, \dots, \boldsymbol{u}_{m_2}$ via (6.4) and calculate the corresponding value of J_1 via (6.3). Then, construct basis vectors

$$L^{p} = \left(\frac{J(e^{p},\mathrm{u},t)}{e^{p}}\right), p = 1, \dots, n-1.$$

And apply CAM to find $\boldsymbol{x}_1{}^*$ based on support vectors $L^{k} = \left(\frac{J(\mathrm{x}^*,\mathrm{u},t)}{x^*}\right)$, where k is the number of iterations for using CAM to solve the problem. Then calculate the corresponding values of other components via (6.4) and $J_1{}^*$.

Next, continue this operation for the state vectors $\boldsymbol{x}_2, \boldsymbol{x}_3, \dots, \boldsymbol{x}_{m_1}$ to find $\boldsymbol{x}_2{}^*, \boldsymbol{x}_3{}^*, \dots, \boldsymbol{x}_{m_1}{}^*$ and corresponding optimal values $J_2{}^*, J_3{}^*, \dots, J_{m_1}{}^*$.

The UVCT initializes 2, 3 and $m_1 - 1$ sate vectors combinations respectively by unit vectors and use CAM to find corresponding optimal values of them in parallel as follows.

CAM can be applied for one state vector or combination of two, three or more states vectors:

$$\boldsymbol{x}_1, \boldsymbol{x}_2; \boldsymbol{x}_1, \boldsymbol{x}_3;\dots; \boldsymbol{x}_1, \boldsymbol{x}_{m_1}; \boldsymbol{x}_2, \boldsymbol{x}_1; \boldsymbol{x}_2, \boldsymbol{x}_3;\dots;$$
$$\boldsymbol{x}_2, \boldsymbol{x}_{m_1},\dots \boldsymbol{x}_1, \boldsymbol{x}_2, \boldsymbol{x}_3;\dots; \boldsymbol{x}_1, \boldsymbol{x}_2, \boldsymbol{x}_{m_1};$$
$$\dots; \boldsymbol{x}_{m_1}, \boldsymbol{x}_{m_1-1}, \dots, \boldsymbol{x}_2$$

and find the corresponding optimal solutions

$$J_{1\,2}{}^{*}, J_{1\,3}{}^{*}, \ldots, J_{1\,m_1}{}^{*}; J_{2\,1}{}^{*}, J_{2\,3}{}^{*}, \ldots, J_{2\,m_1}{}^{*};$$
$$\ldots J_{1\,2\,3}{}^{*}, \ldots, J_{1\,2\,m_1}{}^{*}; \ldots; J_{m_1\,m_1-1\ldots2}{}^{*}$$

respectively. Similarly, CAM can be applied for one control vector or combination of two, three or more control vectors:

$$\boldsymbol{u}_1, \boldsymbol{u}_2, \boldsymbol{u}_3, \ldots, \boldsymbol{u}_{m_1}; \boldsymbol{u}_1, \boldsymbol{u}_2; \boldsymbol{u}_1, \boldsymbol{u}_3;\ldots; \boldsymbol{u}_1, \boldsymbol{u}_{m_1}; \boldsymbol{u}_2, \boldsymbol{u}_1; \boldsymbol{u}_2, \boldsymbol{u}_3;\ldots;$$
$$\boldsymbol{u}_2, \boldsymbol{u}_{m_1},\ldots; \boldsymbol{u}_1, \boldsymbol{u}_2, \boldsymbol{u}_3;\ldots; \boldsymbol{u}_1, \boldsymbol{u}_2, \boldsymbol{u}_{m_1};\ldots;\ldots; \boldsymbol{u}_{m_1}, \boldsymbol{u}_{m_1-1}, \ldots, \boldsymbol{u}_2$$

and find the corresponding optimal solutions

$$\overline{J_1}^{*}, \overline{J_2}^{*}, \overline{J_3}^{*}, \ldots, \overline{J_{m_1}}^{*};$$
$$\overline{J_{1\,2}}^{*}, \overline{J_{1\,3}}^{*}, \ldots, \overline{J_{1\,m_1}}^{*}; \overline{J_{2\,1}}^{*}, \overline{J_{2\,3}}^{*}, \ldots, \overline{J_{2\,m_1}}^{*};$$
$$\ldots; \overline{J_{1\,2\,3}}^{*}, \ldots, \overline{J_{1\,2\,m_1}}^{*}; \ldots; \overline{J_{m_1\,m_1-1\ldots2}}^{*}.$$

The following set is obtained when the CAM is used for the minimization of the discretized OCP:

$$\mathcal{J}^{*} = \{J_1{}^{*}; J_2{}^{*}, J_3{}^{*}, \ldots, J_{m_1}{}^{*}; J_{1\,2}{}^{*}, J_{1\,3}{}^{*}, \ldots, J_{1\,m_1}{}^{*}; J_{2\,1}{}^{*}, J_{2\,3}{}^{*}, \ldots, J_{2\,m_1}{}^{*}; \ldots;$$
$$J_{1\,2\,3}{}^{*}, \ldots, J_{1\,2\,m_1}{}^{*}; \ldots; \ldots; J_{m_1\,m_1-1\ldots2}{}^{*}; \overline{J_1}^{*}, \overline{J_2}^{*}, \overline{J_3}^{*}, \ldots, \overline{J_{m_1}}^{*}$$
$$\overline{J_{1\,2}}^{*}, \overline{J_{1\,3}}^{*}, \ldots, \overline{J_{1\,m_1}}^{*}; \overline{J_{2\,1}}^{*}, \overline{J_{2\,3}}^{*}, \ldots, \overline{J_{2\,m_1}}^{*}; \ldots; \overline{J_{1\,2\,3}}^{*}, \ldots, \overline{J_{1\,2\,m_1}}^{*}; \ldots;$$
$$\ldots; \overline{J_{m_1\,m_1-1\ldots2}}^{*}\}.$$

One can say here is that the smallest element of the set $\mathcal{J}^{*}$ is the global optimum solution of the mentioned OCP based on the global property of CAM.

In some cases, the following suggestions are helpful to decrease the processing time of the algorithm based on OCP assumed conditions.

(i) In some OCP, just use the combinations of the vectors which involve the performance index

(ii) In some examples, just use state vectors or control vectors combinations to converge to the global optimum.

Consider the following global optimization problem:

$$J(\boldsymbol{x}, \boldsymbol{u}, t) \to min$$

$$\text{subject to } \boldsymbol{x}(t), \boldsymbol{u}(t) \in \mathcal{S}^*,$$

where $\mathcal{S}^* = \{t \in \mathbb{R}^n_+ : \sum_i^r \boldsymbol{x}_i(t) = 1, \sum_i^r \boldsymbol{u}_i(t) = 1\}\}$ and r is the number of coordinates of the state and control vectors $\boldsymbol{x}(t), \boldsymbol{u}(t)$.

Let $G: \mathcal{S}^* \to \mathbb{R}$ be a positive Lipschitz functional defined on the unit simplex. Then it can be extended to a finite IPH functional $J(\boldsymbol{x}, \boldsymbol{u}, t)$ on the cone $\mathbb{R}^n_+$, which would coincide with $G(\boldsymbol{x}, \boldsymbol{u}, t) + \mathfrak{c}$ on $\mathcal{S}^*$. I.e., $J(\boldsymbol{x}, \boldsymbol{u}, t) = G(\boldsymbol{x}, \boldsymbol{u}, t) + \mathfrak{c}$ is an IPH functional on $\mathcal{S}^*$. Then, for each $\boldsymbol{x}(t), \boldsymbol{u}(t) \in \mathbb{R}^n_+$ define the support vectors

$$\begin{cases} L_{\boldsymbol{x}} = L = \left(\frac{J(\boldsymbol{x},\boldsymbol{u},t)}{\boldsymbol{x}}\right) = \left(\frac{J(\boldsymbol{x},\boldsymbol{u},t)}{\boldsymbol{x}_1}, \frac{J(\boldsymbol{x},\boldsymbol{u},t)}{\boldsymbol{x}_2}, \dots, \frac{J(\boldsymbol{x},\boldsymbol{u},t)}{\boldsymbol{x}_n}\right) \\ L_{\boldsymbol{u}} = \overline{\overline{L}} = \left(\frac{J(\boldsymbol{x},\boldsymbol{u},t)}{\boldsymbol{u}}\right) = \left(\frac{J(\boldsymbol{x},\boldsymbol{u},t)}{\boldsymbol{u}_1}, \frac{J(\boldsymbol{x},\boldsymbol{u},t)}{\boldsymbol{u}_2}, \dots, \frac{J(\boldsymbol{x},\boldsymbol{u},t)}{\boldsymbol{u}_n}\right) \end{cases}.$$

The n vectors $e^m = (0, \dots, 0,1,0, \dots, 0)$, with 1 in the mth position shall be used, and the corresponding support vectors $\begin{cases} L^m = \left(\frac{J(e^m,\boldsymbol{u},t)}{e^m}\right) \\ \overline{\overline{L}}^m = \left(\frac{J(\boldsymbol{x},e^m,t)}{e^m}\right) \end{cases}$, $m = 1, \dots, n$ shall be called basis vectors. A set of $k \geq n$ support vectors (and hence k known values of the functional $J(\boldsymbol{x}, \boldsymbol{u}, t)$ at k distinct vectors), $\begin{cases} \mathcal{K}_{\boldsymbol{x}} = \{L^k\}_{j=1}^k \\ \mathcal{K}_{\boldsymbol{u}} = \{\overline{\overline{L}}^k\}_{j=1}^k \end{cases}$. Let also the first n support vectors be the basis vectors. This choice of support vectors guarantees that the algorithm will locate all local (and hence global) minimizers of the auxiliary functionals $\begin{cases} H_{k}(\boldsymbol{x},\boldsymbol{u},t) = \max\limits_{j\geq k} \min\limits_{i=1,\dots,n} L_i^j \boldsymbol{x}_i \\ \overline{\overline{H}}_{k}(\boldsymbol{x},\boldsymbol{u},t) = \max\limits_{j\geq k} \min\limits_{i=1,\dots,n} \overline{\overline{L}}_i^j \boldsymbol{u}_i \end{cases}$ in the interior of the unit simplex. It always underestimates the value of $J(\boldsymbol{x},\boldsymbol{u},t)$: $\begin{cases} H_{k}(\boldsymbol{x}) \leq J(\boldsymbol{x},\boldsymbol{u},t) \\ \overline{\overline{H}}_{k}(\boldsymbol{x}) \leq J(\boldsymbol{x},\boldsymbol{u},t) \end{cases}$. Hence, $\begin{cases} b_{k} = \min\limits_{x\in S^*} H_{k}(\boldsymbol{x},\boldsymbol{u},t) \leq \min\limits_{x\in S^*} J(\boldsymbol{x},\boldsymbol{u},t) \\ \overline{\overline{b}}_{k} = \min\limits_{x\in S^*} \overline{\overline{H}}_{k}(\boldsymbol{x},\boldsymbol{u},t) \leq \min\limits_{u\in S^*} J(\boldsymbol{x},\boldsymbol{u},t) \end{cases}$.

On the otherhand, the sequence of its minima, $\begin{cases} \{b_{k}\}_{k=n}^{\infty} \\ \{\overline{\overline{b}}_{k}\}_{k=n}^{\infty} \end{cases}$ is increasing, and convergence to the global minmum of $J(\boldsymbol{x},\boldsymbol{u},t)$ based on its function framework which presented by Rubinov (2000), Bagirov and Rubinov (2000). And based the CAM was formulated by Rubinov (2000) on function the following algorithms are presented for functional in general (or for OCP in particular).

6.2.1. The CAM for Functional

Algorithm 1

Step 0. Initialization

(a) Take points e^m, $m = 1, \dots, n$, and construct basic vectors $L^m = (\frac{J(e^m,\boldsymbol{u},t)}{e^m})$, $m = 1, \dots, n$.

(b) Define the functional $H_n(x) = \max\limits_{j\leq n} \min\limits_{i=1,\dots,n} L_i^j x_i = \max\limits_{j\leq n} L_j^j x_j$.

(c) Set $k = n$.

Step 1. Find $x^* = \arg[\min\limits_{x\in S^*} H_{k}(x)]$.

Step 2. Set $k = k + 1$ and $x^k = x^*$.

Step 3. Compute $L^j = \left(\frac{J(x^k,\boldsymbol{u},t)}{x^k}\right)$. Define the function

$$H_{k}(x) = \max_{j\leq k} \min_{i=1,\dots,n} L_i^j x_i = \max\{H_{k-1}(x), \min_{i=1,\dots,n} L_i^k x_i\}.$$

Go to Step 2.

The most time-consuming steps of the algorithms 1 and 2 are their Step 1, minimization of the auxiliary functional. The next theorem will improve this problem.

Theorem 6.1: *Let* $\boldsymbol{x}^*(t) > 0$ (or $\boldsymbol{u}^*(t) > 0$) *be a local minimizer of* $H_k(\boldsymbol{x},\boldsymbol{u},t)$ (or $\overline{\overline{H}}_k(\boldsymbol{x},\boldsymbol{u},t)$) *over the relative interior of* $\mathcal{S}^*$, $ri\mathcal{S}^* = \{\boldsymbol{x}(t) \in \mathcal{S}^*, \boldsymbol{x}(t) > 0\}$ (or $ri\mathcal{S}^* = \{\boldsymbol{u}(t) \in \mathcal{S}^*, \boldsymbol{u}(t) > 0\}$). *Then there exists a subset* $\mathcal{L} = \{L^{j_1}, L^{j_2}, \dots, L^{j_n}\}$ (or $\overline{\overline{\mathcal{L}}} = \{\overline{\overline{L}}^{j_1}, \overline{\overline{L}}^{j_2}, \dots, \overline{\overline{L}}^{j_n}\}$) *of the set* $\mathcal{K}$ (or $\overline{\overline{\mathcal{K}}}$), *such that*

1. $\boldsymbol{x} = \{d/L_1^{j_1}, d/L_2^{j_2}, \dots, d/L_n^{j_n}\}$ with $d = (\sum_i 1/L^{j_i})^{-1}$
   (or $\boldsymbol{u} = \{d/\overline{\overline{L}}_1^{j_1}, d/\overline{\overline{L}}_2^{j_2}, \dots, d/\overline{\overline{L}}_n^{j_n}\}$ with $d = (\sum_i 1/\overline{\overline{L}}^{j_i})^{-1}$).
2. $\max_{j \le k} \min_{i=1,\dots,n} L_i^j / L_i^{j_i} = 1$
   (or $\max_{j \le k} \min_{i=1,\dots,n} \overline{\overline{L}}_i^j / \overline{\overline{L}}_i^{j_i} = 1$.
3. *Either* $\forall i: j_i = i$, *or* $\exists m: k_m > n, L_i^{j_m} > L_i^{j_i}, \forall i \ne m$
   (or *Either* $\forall i: j_i = i$, *or* $\exists m: k_m > n, \overline{\overline{L}}_i^{j_m} > \overline{\overline{L}}_i^{j_i}, \forall i \ne m$.).

Proof: Completely similar to the proof on a function by Bagirov and Rubinov (2003).

The value of the auxiliary functional at $\boldsymbol{x}$ (or $\boldsymbol{u}$) is

$$H_k(\boldsymbol{x},\boldsymbol{u},t) = d \text{ (or } \overline{\overline{H}}_k(\boldsymbol{x},\boldsymbol{u},t) = d).$$

The following version of CAM for functionals works based on the above theorem, by examining all possible combinations of n support vectors.

Consider a set of k support vectors $\mathcal{K} = \{L^j\}_{j=1}^k$ (or $\overline{\overline{\mathcal{K}}} = \{\overline{\overline{L}}^j\}_{j=1}^k$), $L^j \in \mathbb{R}_+^n$ (or $\overline{\overline{L}}^j \in \mathbb{R}_+^n$). Let $\mathcal{J}$ denote $\{1,2,\dots,n\}$. From Theorem 6.1, the local minima of the auxiliary functional $H_{k}(\boldsymbol{x},\boldsymbol{u},t)$ (or $\bar{\bar{H}}_{k}(\boldsymbol{x},\boldsymbol{u},t)$ are combinations of n support vectors $\mathcal{L} = \{L^{j_1}, L^{j_2}, \dots, L^{j_n}\}$ (or $\bar{\bar{\mathcal{L}}} = \{\bar{\bar{L}}^{j_1}, \bar{\bar{L}}^{j_2}, \dots, \bar{\bar{L}}^{j_n}\}$) that satisfy the following conditions:

(I) $\forall i, s \in \mathcal{I}, i \neq s: L_i^{j_i} < L_i^{j_s}$ (or $\bar{\bar{L}}_i^{j_i} < \bar{\bar{L}}_i^{j_s}$)

(II) $\forall v \in \mathcal{K} \backslash \mathcal{L}, \exists i \in \mathcal{I}: L_i^{j_i} \geq v_i$ (or $\forall \bar{\bar{v}} \in \bar{\bar{\mathcal{K}}} \backslash \bar{\bar{\mathcal{L}}}, \exists i \in \mathcal{I}: \bar{\bar{L}}_i^{j_i} \geq \bar{\bar{v}}_i$).

The subset $\mathcal{L}$ (or $\bar{\bar{\mathcal{L}}}$), which satisfied conditions (I) and (II) above, is called a valid combination of support vectors.

Let $\mathcal{W}^{k}$ (or $\bar{\bar{\mathcal{W}}}^{k}$) denote the set of all valid combinations $\mathcal{L}$ (or $\bar{\bar{\mathcal{L}}}$) of k support vectors satisfying conditions (I) and (II):

$$\mathcal{W}^{k} = \{\mathcal{L} = \{L^{j_1}, L^{j_2}, \dots, L^{j_n}\}, L^{j_i} \in \mathcal{K}: (1) \forall i, s \in \mathcal{I}, i \neq s: \mathcal{L}_{ii} < \mathcal{L}_{si} \text{ and } (2)\ \forall v \in \mathcal{K} \backslash \mathcal{L}\ \exists i \in \mathcal{I}: \mathcal{L}_{ii} \geq v_i\}$$

$$(\text{or } \bar{\bar{\mathcal{W}}}^{k} = \{\bar{\bar{\mathcal{L}}} = \{\bar{\bar{L}}^{j_1}, \bar{\bar{L}}^{j_2}, \dots, \bar{\bar{L}}^{j_n}\}, \bar{\bar{L}}^{j_i} \in \bar{\bar{\mathcal{K}}}: (1) \forall i, s \in \mathcal{I}, i \neq s: \bar{\bar{\mathcal{L}}}_{ii} < \bar{\bar{\mathcal{L}}}_{si} \text{ and } (2)\ \forall \bar{\bar{v}} \in \bar{\bar{\mathcal{K}}} \backslash \bar{\bar{\mathcal{L}}}\ \exists i \in \mathcal{I}: \bar{\bar{\mathcal{L}}}_{ii} \geq \bar{\bar{v}}_i\})$$

The problem of finding local minima of $H_{k}(\boldsymbol{x},\boldsymbol{u},t)$ (or $\bar{\bar{H}}_{k}(\boldsymbol{x},\boldsymbol{u},t)$ is translated into the problem of listing the elements of $\mathcal{W}^{k}$ (or $\bar{\bar{\mathcal{W}}}^{k}$). A simplistic approach is then:

Step 1. Construct all combinations $\mathcal{L}$ (or $\bar{\bar{\mathcal{L}}}$) satisfying (I)
Step 2. Check the obtained combinations against (II).

Then improve on this by taking into account the fact that all elements of $\mathcal{W}^{k}$ (or $\bar{\bar{\mathcal{W}}}^{k}$) that do not involve L^{j} (or $\bar{\bar{L}}^{j}$) (i.e., minima of

$H_{k-1}(\boldsymbol{x},\boldsymbol{u},t)$ (or $\overline{\overline{H}}_{k-1}(\boldsymbol{x},\boldsymbol{u},t)$) do not need to be recomputed, hence the algorithm takes the form:

Algorithm 2

Step 0. Initialization

(a) Take points e^m, $m = 1, \dots, n$, and construct basic vectors $L^m = \left(\frac{J(e^m,\boldsymbol{u},t)}{e^m}\right)$, $m = 1, \dots, n$.

(b) Define the function $H_n(\boldsymbol{x},\boldsymbol{u},t) = \max_{j \le n} \min_{i=1,\dots,n} L_i^j \boldsymbol{x}_i = \max_{j \le n} L_j^j \boldsymbol{x}_j$.

(c) Set $k = n$. Set $\mathcal{W}^k = \{\{L^1, L^2, \dots, L^n\}\}$.

(d) Calculate $d = (\sum_{i=1,\dots n} 1/L_i^i)^{-1}$.

Step 1.

(a) Retrieve all valid combinations $\mathcal{L}$ (i.e., $\mathcal{W}^k$).

(b) Select $\mathcal{L} \in \mathcal{W}^k$.

Step 2.

(a) $k = k + 1$.

(b) Take $\boldsymbol{x}^* = d/\mathrm{diag}(\mathcal{L})$ and evaluate $J(\boldsymbol{x}^*,\boldsymbol{u},t)$.

(c) Compute $L^k = (J(\boldsymbol{x}^*,\boldsymbol{u},t)/x^*)$. Define the function

$$H_k(\boldsymbol{x},\boldsymbol{u},t) = \max_{j \le k} \min_{i=1,\dots,n} L_i^j \boldsymbol{x}_i = \max\{H_{k-1}(\boldsymbol{x}), \min_{i=1,\dots,n} L_i^k \boldsymbol{x}_i\}$$

Step 3.

(a) Check $\mathcal{W}^{k-1}$ against (II) and remove those that fail (II).

(b) Move the remaining combinations into $\mathcal{W}^k$.

Step 4.

(a) Construct all combinations $\mathcal{L}$ that involve L^{k} and satisfy (I).
(b) Calculate $d = (\sum_{i=1,\dots n} 1/L_i^{k_i})^{-1}$ for each such combination.
(c) Add these combinations to $\mathcal{W}^{k}$.
(d) Go to Step 1.

Remark 6.1: In Algorithm 2 if the diagonal of $\mathcal{L}$ (or $\bar{\bar{\mathcal{L}}}$) was dominated by L^{k}, this combination will be not moved into $\mathcal{W}^{k+1}$.

Remark 6.2: In Algorithm 2, each iteration the algorithm computes and checks the value of $J(\boldsymbol{x}^*, \boldsymbol{u}, t) - d$ (or $J(\boldsymbol{x}, \boldsymbol{u}^*, t) - d$) with the value of assumed stopping criteria. While it satisfies then the algorithm stopped.

It is clear that the set $\mathcal{J}^*$ is countably finite. Furthermore, with regards to Proposition 6.1 obtained based on Theorems 6.4 and 6.5, $\mathcal{J}^*$ is not empty.

Theorem 6.4: *Suppose that the function f is continuous on the closed interval $[a, d]$. If $\mathfrak{m}$ and $\mathfrak{n}$ are, respectively, the absolute minimum and absolute maximum function values of f on $[a, d]$ so that [4],*

$$\mathfrak{m} \le f(x) \le \mathfrak{n} \text{ for } a \le x \le d$$

then

$$\mathfrak{m}(d - a) \le \int_a^d f(x)\,dx \le \mathfrak{n}(d - a).$$

Theorem 6.5 (The Mean-Value Theorem for Integrals): *If f is continuous on the closed interval $[a, d]$, there exists a number c in $[a, d]$ such that,*

$$\int_a^d f(x)\,dx = f(c)(d-a).$$

Proposition 6.1: The set of all optimal solutions, $\mathcal{J}^*$, by CAM for discretized OCP is not empty.

Proof: the original OCP can be shown as follows

$$\begin{aligned} J(\boldsymbol{x},\boldsymbol{u},t) &= \int_a^d E(\boldsymbol{x},\boldsymbol{u},t)\,dt \\ &= \int_a^d F_1(\boldsymbol{x},\boldsymbol{u},t)\,dt + \int_a^d F_2(\boldsymbol{x},\boldsymbol{u},t)\,dt + \cdots \\ &+ \int_a^d F_n(\boldsymbol{x},\boldsymbol{u},t)\,dt \end{aligned}$$

where $E(\boldsymbol{x},\boldsymbol{u},t) = F_1(\boldsymbol{x},\boldsymbol{u},t) + F_2(\boldsymbol{x},\boldsymbol{u},t) + \cdots + F_n(\boldsymbol{x},\boldsymbol{u},t)$.

Based on Theorems 6.4 and 6.5 the following assertions shall be presented

$$\mathfrak{m}_1(d-a) \le \int_a^d F_1(\boldsymbol{x},\boldsymbol{u},t)\,dt = F_1(c_1)(d-a)$$

$$\mathfrak{m}_2(d-a) \le \int_a^d F_2(\boldsymbol{x},\boldsymbol{u},t)\,dt = F_2(c_2)(d-a)$$

$$\vdots$$

$$\mathfrak{m}_n(d-a) \le \int_a^d F_n(\boldsymbol{x},\boldsymbol{u},t)\,dt = F_n(c_n)(d-a) \tag{6.5}$$

Next, (6.5) implies that

$$\mathfrak{m}_1(d-a)+\mathfrak{m}_2(d-a)+\cdots+\mathfrak{m}_n(d-a)$$
$$\leq \int_a^d F_1(\boldsymbol{x},\boldsymbol{u},t)\,dt+\int_a^d F_2(\boldsymbol{x},\boldsymbol{u},t)\,dt+\cdots$$
$$+\int_a^d F_n(\boldsymbol{x},\boldsymbol{u},t)\,dt$$
$$=F_1(c_1)(d-a)+F_2(c_2)(d-a)+\cdots+F_n(c_n)(d-a)$$

$$\Rightarrow (\mathfrak{m}_1+\mathfrak{m}_2+\cdots+\mathfrak{m}_n)(d-a)\leq J(\boldsymbol{x},\boldsymbol{u},t)=c(d-a)$$
$$\Rightarrow \mathfrak{m}(d-a)\leq J(\boldsymbol{x},\boldsymbol{u},t)=c(d-a)$$

where $\mathfrak{m}=\mathfrak{m}_1+\mathfrak{m}_2+\cdots+\mathfrak{m}_n$ and $c=F_1(c_1)+F_2(c_2)+\cdots+F_n(c_n)$

Thus

$$\mathfrak{m}\leq c$$

Then there exists a $\mathcal{q}\in\mathbb{R}$ such that $\mathcal{q}=\mathfrak{m}$ or $\mathcal{q}=c$ or $\mathfrak{m}<\mathcal{q}<c$. Then implies that

$$J_{\mathcal{q}}\leq J(\boldsymbol{x},\boldsymbol{u},t)\Rightarrow J_{\mathcal{q}}\in\mathcal{J}.$$

where $\mathcal{q}(d-a)\leq J_{\mathcal{q}}=J(\boldsymbol{x}_{\mathcal{q}},\boldsymbol{u}_{\mathcal{q}},t)$ or $J_{\mathcal{q}}=J\left(\boldsymbol{x}_{\mathcal{q}},\boldsymbol{u}_{\mathcal{q}},t\right)=\mathcal{q}(d-a)$.

Thus $J_{\mathcal{q}}\in\mathcal{J}^*$ and the set $\mathcal{J}^*$ is not empty ■.

Example 6.1: The first minimization test on OCP is

$$J=\int_0^1(\boldsymbol{x}^2(t)+\boldsymbol{u}(t))dt$$

$$s.t\ \dot{\boldsymbol{x}}=\boldsymbol{u}(t)$$

$$\boldsymbol{x}(0)=0,\boldsymbol{x}(1)=0.5$$

The discretized form of above OCP is

$$J = \mathfrak{h} \sum_{i=1}^{n-1} (\boldsymbol{x}^2(t) + \boldsymbol{u}(t))$$

$$s.t\ \boldsymbol{x} = \mathfrak{h} * \boldsymbol{u}(t)$$

$$\boldsymbol{x}(0) = 0, \boldsymbol{x}(1) = 0.5,$$

where $\boldsymbol{x}(t)$ and $\boldsymbol{u}(t)$ are state and control vectors respectively.

For this problem, the interval $[0, 1]$ was partitioned into 10 and 20 equal subintervals. Then, for $n = 10$: $t_1 = 0, t_2 = 0.1, \dots, t_{10} = 1$ and for $n = 20$: $t_1 = 0, t_2 = 0.05, \dots, t_{20} = 1$, the CAM is applied with three iterations to minimize the mentioned OCP using both partitions. The results show that the number of partitions is an efficient factor for finding the global optimal solution with stopping criterion= 0.001.

For $n = 10$: $t_1 = 0, t_2 = 0.1, \dots, t_{10} = 1$, the step size $\hbar = \frac{1-0}{10} = 0.1$ and based on Procedure 6.1, the discretized OCP becomes:

$$J = 0.1 \sum_{i=1}^{9} (\boldsymbol{x}^2(t) + \boldsymbol{u}(t))$$

$$s.t\ \boldsymbol{x} = 0.1 * \boldsymbol{u}(t)$$

$$\boldsymbol{x}(0) = 0, \boldsymbol{x}(1) = 0.5,$$

Firstly, the CAM is used to find $\boldsymbol{x}^*$,

Step 0. Initialization

Take points $e^1 = (1,0,0, \dots, 0)_{1\times 9}, e^2 = (0,1,0, \dots, 0)_{1\times 9}, \dots, e^9 = (0, \dots, 0,1)_{1\times 9}$.

Compute J_i $(i = 1, \dots, 9)$:

$$\boldsymbol{u}_1 = 10 * \boldsymbol{x}_1 = 10 * e^1 = (10,0,0,\dots,0)_{1\times9}, \dots,$$

$$\boldsymbol{u}_9 = 10 * \boldsymbol{x}_9 = 10 * e^9 = (0,\dots,0,10)_{1\times9}$$

$$J_1(e^1, \mathrm{u}_1, t) = 0.11, J_2(e^2, \mathrm{u}_2, t) = 0.22, \dots, J_9(e^9, \mathrm{u}_9, t) = 0.11$$

and construct basis vectors,

$$L^1 = \left(\frac{J_1(e^1, \mathrm{u}, t)}{e^1}\right) = (\frac{0.11}{1}, \frac{0.11}{0}, \frac{0.11}{0}, \dots, \frac{0.11}{0})_{1\times9} = (0.11, \infty, \infty, \dots, \infty)_{1\times9}$$

$$L^2 = \left(\frac{J_1(e^2, \mathrm{u}, t)}{e^2}\right) = (\frac{0.22}{1}, \frac{0.22}{0}, \frac{0.22}{0}, \dots, \frac{0.22}{0})_{1\times9} = (\infty, 0.22, \infty, \dots, \infty)_{1\times9}$$

$$L^3 = \left(\frac{J_1(e^3, \mathrm{u}, t)}{e^3}\right) = (\frac{0.33}{1}, \frac{0.33}{0}, \frac{0.33}{0}, \dots, \frac{0.33}{0})_{1\times9} = (\infty, \infty, 0.33, \infty, \dots, \infty)_{1\times9}$$

$$L^4 = \left(\frac{J_1(e^4, \mathrm{u}, t)}{e^4}\right) = (\frac{0.22}{0}, \dots, \frac{0.22}{1}, \dots, \frac{0.22}{0})_{1\times9} = (\infty, \infty, \infty, 0.22, \infty, \dots, \infty)_{1\times9}$$

$$L^5 = \left(\frac{J_1(e^5, \mathrm{u}, t)}{e^5}\right) = (\frac{0.22}{0}, \dots, \frac{0.22}{1}, \dots, \frac{0.22}{0})_{1\times9} = (\infty, \dots, \infty, 0.22, \infty, \dots, \infty)_{1\times9}$$

$$L^6 = \left(\frac{J_1(e^6, \mathrm{u}, t)}{e^6}\right) = (\frac{0.22}{0}, \dots, \frac{0.22}{1}, \dots, \frac{0.22}{0})_{1\times9} = (\infty, \dots, \infty, 0.22, \infty, \dots, \infty)_{1\times9}$$

$$L^7 = \left(\frac{J_1(e^7, \mathrm{u}, t)}{e^7}\right) = (\frac{0.22}{0}, \dots, \frac{0.22}{1}, \dots, \frac{0.22}{0})_{1\times 9}$$
$$= (\infty, \dots, \infty, 0.22, \infty, \infty, \infty)_{1\times 9}$$

$$L^8 = \left(\frac{J_1(e^8, \mathrm{u}, t)}{e^8}\right) = (\frac{0.11}{0}, \dots, \frac{0.11}{1}, \frac{0.11}{0})_{1\times 9} = (\infty, \dots, 0.11, \infty)_{1\times 9}$$

$$L^9 = \left(\frac{J_1(e^9, \mathrm{u}, t)}{e^9}\right) = (\frac{0.11}{1}, \frac{0.11}{0}, \frac{0.11}{0}, \dots, \frac{0.11}{0})_{1\times 9}$$
$$= (\infty, \infty, \dots, 0.11)_{1\times 9}$$

Set $k = 9$ and $\mathcal{W}^9 = \left\{\{L^1, L^2, L^3, L^4, L^5, L^6, L^7, L^8, L^9\}\right\} = \begin{pmatrix} 0.11 & \cdots & \infty \\ \vdots & \ddots & \vdots \\ \infty & \cdots & 0.11 \end{pmatrix}$

Calculate d,

$$d = \frac{1}{\Sigma(\frac{1}{L_1^1} + \frac{1}{L_2^2} + \cdots + \frac{1}{L_9^9})} = 0.0189$$

A local minimum d can be found from the combination of basis vectors in initialization where d is the value of a local minimum of auxiliary functional H_9^1. Now the Iteration 1 is:

Iteration 1

Step 1. Select $\{L^1, L^2, L^3, L^4, L^5, L^6, L^7, L^8, L^9\}$ since it is the only element in $\mathcal{W}^9$.

Step 2. Set $k = 10$.

Take

$$\boldsymbol{x}^* = \frac{d}{diag(\{L^1, L^2, L^3, L^4, L^5, L^6, L^7, L^8, L^9\})}$$
$$= \left(\frac{0.0189}{0.11}, \frac{0.0189}{0.22}, \ldots, \frac{0.0189}{0.11}\right)$$

$$= (0.1714, 0.0857, 0.0571, 0.0857, 0.0857, 0.0857, 0.0857, 0.1714, 0.1714)$$

Evaluate $J_1^*(\boldsymbol{x}^*)$

$$J_1^*(\boldsymbol{x}^*) = 0.0228$$
$$Criteria = J_1^*(\boldsymbol{x}^*) - d = 0.004$$

Compute L^{10},

$$L^{10} = \left(\frac{J(\boldsymbol{x}^*, \boldsymbol{u}, t)}{\boldsymbol{x}^*}\right)$$

$$= (0.1331\ 0.2662\ 0.3993\ 0.2662\ 0.2662\ 0.2662\ 0.2662\ 0.1331\ 0.1331)$$

Step 3. Check $\mathcal{W}^9$ against conditions (2): for every vector $\boldsymbol{v}$ of $\mathcal{K} = \{L^1, L^2, L^3, L^4, L^5, L^6, L^7, L^8, L^9, L^{10}\}$ which is taken, that not already in L, the diagonal of L is not dominated by v.

Since diagonal of $\{L^1, L^2, L^3, L^4, L^5, L^6, L^7, L^8, L^9\}$ is dominated by L^{10}, this combination will be not moved into $\mathcal{W}^{10}$.

Step 4. Construct all combinations L that involve L^{10} and satisfy condition (1): every element on the diagonal must be the smallest in its column.

Combinations that involve L^{10}:

$$\{L^{10}, L^2, L^3, L^4, L^5, L^6, L^7, L^8, L^9\}, \{L^1, L^{10}, L^3, L^4, L^5, L^6, L^7, L^8, L^9\},$$
$$\{L^1, L^2, L^{10}, L^4, L^5, L^6, L^7, L^8, L^9\}, \{L^1, L^2, L^3, L^{10}, L^5, L^6, L^7, L^8, L^9\},$$
$$\{L^1, L^2, L^3, L^4, L^{10}, L^6, L^7, L^8, L^9\}, \{L^1, L^2, L^3, L^4, L^5, L^{10}, L^7, L^8, L^9\},$$
$$\{L^1, L^2, L^3, L^4, L^5, L^6, L^{10}, L^8, L^9\}, \{L^1, L^2, L^3, L^4, L^5, L^6, L^7, L^{10}, L^9\},$$
$$\{L^1, L^2, L^3, L^4, L^5, L^6, L^7, L^8, L^{10}\}$$

Combinations satisfying condition (1) and their value of d,

$$d = \frac{1}{\Sigma(\frac{1}{L_1^{k_1}} + \frac{1}{L_2^{k_2}} + \cdots + \frac{1}{L_9^{k_9}})}$$

$$\{L^{10}, L^2, L^3, L^4, L^5, L^6, L^7, L^8, L^9\},\ d = 0.0194$$
$$\{L^1, L^{10}, L^3, L^4, L^5, L^6, L^7, L^8, L^9\},\ d = 0.0191$$
$$\{L^1, L^2, L^{10}, L^4, L^5, L^6, L^7, L^8, L^9\},\ d = 0.0198$$
$$\{L^1, L^2, L^3, L^{10}, L^5, L^6, L^7, L^8, L^9\},\ d = 0.0202$$
$$\{L^1, L^2, L^3, L^4, L^{10}, L^6, L^7, L^8, L^9\},\ d = 0.0202$$
$$\{L^1, L^2, L^3, L^4, L^5, L^{10}, L^7, L^8, L^9\},\ d = 0.0202$$
$$\{L^1, L^2, L^3, L^4, L^5, L^6, L^{10}, L^8, L^9\},\ d = 0.0202$$
$$\{L^1, L^2, L^3, L^4, L^5, L^6, L^7, L^{10}, L^9\},\ d = 0.0205$$
$$\{L^1, L^2, L^3, L^4, L^5, L^6, L^7, L^8, L^{10}\},\ d = 0.0205$$

Add these combinations to $\mathcal{W}^{10}$. Therefore, we have:

$$\mathcal{W}^{10} =$$
$$\{\{L^{10}, L^2, L^3, L^4, L^5, L^6, L^7, L^8, L^9\}, \{L^1, L^{10}, L^3, L^4, L^5, L^6, L^7, L^8, L^9\},$$
$$\{L^1, L^2, L^{10}, L^4, L^5, L^6, L^7, L^8, L^9\}, \{L^1, L^2, L^3, L^{10}, L^5, L^6, L^7, L^8, L^9\},$$
$$\{L^1, L^2, L^3, L^4, L^{10}, L^6, L^7, L^8, L^9\}, \{L^1, L^2, L^3, L^4, L^5, L^{10}, L^7, L^8, L^9\},$$
$$\{L^1, L^2, L^3, L^4, L^5, L^6, L^{10}, L^8, L^9\}, \{L^1, L^2, L^3, L^4, L^5, L^6, L^7, L^{10}, L^9\},$$
$$\{L^1, L^2, L^3, L^4, L^5, L^6, L^7, L^8, L^{10}\}\}$$

End of iteration 1.

After evaluating the value of $J_1(\boldsymbol{x}^*, \boldsymbol{u}, \boldsymbol{t})$, the stopping criterion is checked, the value of $J_1(\boldsymbol{x}^*, \boldsymbol{u}, \boldsymbol{t})$ is the upper estimate and d is the lower estimate of global minimum. If the difference between the two values is less than a given tolerance, the iteration is stopped. If it is not less than a given tolerance, we proceed by creating a new support vector, L^{10}. Since a new support vector is added, the auxiliary functional is modified and become H_{10}^{1}. One can see here is some of the local minima may disappear and the new ones will appear. In Step 3 we check whether a local minimum of H_{9}^{1} is also a local minimum of H_{10}^{1}. In Step 4, new combinations are created. If the new combination satisfies condition (1) then it will give a local minima so the combination will be added into set $\mathcal{W}^{10}$.

Iteration 2

Step 1. Select combination in $\mathcal{W}^{10}$ with the smallest d: $\{L^1, L^{10}, L^3, L^4, L^5, L^6, L^7, L^8, L^9\}$.

Step 2. Set $k = 11$.

and take

$$\boldsymbol{x}^* = \frac{d}{diag(\{L^1, L^2, L^3, L^4, L^5, L^6, L^7, L^8, L^{10}\})}$$
$$= \left(\frac{0.0194}{0.11}, \frac{0.0194}{0.22}, \dots, \frac{0.0194}{0.1331}\right)$$

$$= (0.1767, 0.0883, 0.0589, 0.0883, 0.0883, 0.0883, 0.0883, 0.1767, 0.1460)$$

Evaluate $J_1(\boldsymbol{x}^*, \boldsymbol{u}, t)$,

$$J_1(\boldsymbol{x}^*, \boldsymbol{u}, t) = 0.0226.$$

$$J_1(\boldsymbol{x}^*, \boldsymbol{u}, t) - d = 0.0032$$

Compute L^{11},

$$L^{11} = \left(\frac{J(\boldsymbol{x}^*, \boldsymbol{u}, t)}{\boldsymbol{x}^*}\right)$$
$$= (0.1281, 0.2561, 0.3842, 0.2561, 0.2561, 0.2561, 0.2561, 0.1281, 0.1549)$$

Step 3. Check $\mathcal{W}^{10}$ against condition (2): for every vector v of $\mathcal{K} = \{L^1, L^2, L^3, L^4, L^5, L^6, L^7, L^8, L^{10}, L^{11}\}$ is taken, that not already in $\mathcal{L}$, the diagonal of $\mathcal{L}$ is not dominated by $\mathcal{v}$.

Since diagonal of $\{L^1, L^2, L^3, L^4, L^5, L^6, L^7, L^8, L^{10}\}$ is dominated by L^{11}, this combination will not be moved into $\mathcal{W}^{11}$

Step 4. Construct all combinations L that involve L^{11} and satisfy condition (1): every element on the diagonal must be the smallest in its column.

Combinations that involve L^{11}:

$$\{L^{11}, L^2, L^3, L^4, L^5, L^6, L^7, L^8, L^{10}\}, \{L^1, L^{11}, L^3, L^4, L^5, L^6, L^7, L^8, L^{10}\}$$
$$\{L^1, L^2, L^{11}, L^4, L^5, L^6, L^7, L^8, L^{10}\}, \{L^1, L^2, L^3, L^{11}, L^5, L^6, L^7, L^8, L^{10}\}$$
$$\{L^1, L^2, L^3, L^4, L^{11}, L^6, L^7, L^8, L^{10}\}, \{L^1, L^2, L^3, L^4, L^5, L^{11}, L^7, L^8, L^{10}\}$$
$$\{L^1, L^2, L^3, L^4, L^5, L^6, L^{11}, L^8, L^{10}\}, \{L^1, L^2, L^3, L^4, L^5, L^6, L^7 L^{11}, L^{10}\}$$
$$\{L^1, L^2, L^3, L^4, L^5, L^6, L^7, L^8, L^{11}\}$$

Combinations satisfying condition (1) and their value of d is:

$$d = \frac{1}{\Sigma(\frac{1}{L_1^{k_1}} + \frac{1}{L_2^{k_2}} + \cdots + \frac{1}{L_9^{k_9}})}$$

$$\{L^{11}, L^2, L^3, L^4, L^5, L^6, L^7, L^8, L^{10}\}, d = 0.0199$$
$$\{L^1, L^{11}, L^3, L^4, L^5, L^6, L^7, L^8, L^{10}\}, d = 0.0198$$
$$\{L^1, L^2, L^{11}, L^4, L^5, L^6, L^7, L^8, L^{10}\}, d = 0.0205$$
$$\{L^1, L^2, L^3, L^{11}, L^5, L^6, L^7, L^8, L^{10}\}, d = 0.0208$$
$$\{L^1, L^2, L^3, L^4, L^{11}, L^6, L^7, L^8, L^{10}\}, d = 0.0208$$
$$\{L^1, L^2, L^3, L^4, L^5, L^{11}, L^7, L^8, L^{10}\}, d = 0.0208$$
$$\{L^1, L^2, L^3, L^4, L^5, L^6, L^{11}, L^8, L^{10}\}, d = 0.0208$$
$$\{L^1, L^2, L^3, L^4, L^5, L^6, L^7 L^{11}, L^{10}\}, d = 0.0211$$
$$\{L^1, L^2, L^3, L^4, L^5, L^6, L^7, L^8, L^{11}\}, d = 0.0210$$

Add these combinations to $\mathcal{W}^{11}$. Thus, we have that

$$\begin{aligned}\mathcal{W}^{11} = \{&\{L^{11}, L^2, L^3, L^4, L^5, L^6, L^7, L^8, L^{10}\},\\ &\{L^1, L^{11}, L^3, L^4, L^5, L^6, L^7, L^8, L^{10}\}\\ &\{L^1, L^2, L^{11}, L^4, L^5, L^6, L^7, L^8, L^{10}\},\\ &\{L^1, L^2, L^3, L^{11}, L^5, L^6, L^7, L^8, L^{10}\}\\ &\{L^1, L^2, L^3, L^4, L^{11}, L^6, L^7, L^8, L^{10}\},\\ &\{L^1, L^2, L^3, L^4, L^5, L^{11}, L^7, L^8, L^{10}\}\\ &\{L^1, L^2, L^3, L^4, L^5, L^6, L^{11}, L^8, L^{10}\},\\ &\{L^1, L^2, L^3, L^4, L^5, L^6, L^7 L^{11}, L^{10}\}\\ &\{L^1, L^2, L^3, L^4, L^5, L^6, L^7, L^8, L^{11}\}\}\end{aligned}$$

End of iteration 2.

After evaluating the value of $J_1(\boldsymbol{x}^*, \boldsymbol{u}, t)$, the stopping criterion is checked, the value of $J_1(\boldsymbol{x}^*, \boldsymbol{u}, t)$ is the upper estimate and d is the lower estimate of global minimum. If the difference between the two values is less than a given tolerance, the iteration is stopped. If it is not less than a given tolerance, create a new support vector, L^{11}. Since a new support vector is added, the auxiliary functional is modified and become H_{11}^1. Step 3 checks whether a local minimum of H_{10}^1 is also a local minimum of H_{11}^1. In Step 4, new combinations are created. If the new combination satisfies condition (1) then it will give a local minimum. Thus, the combination is added into set $\mathcal{W}^{11}$.

Iteration 3

Step 1. Select combination in $\mathcal{W}^{11}$ with the smallest d, which is $\{L^1, L^{11}, L^3, L^4, L^5, L^6, L^7, L^8, L^{10}\}$.

Step 2. Set $k = 12$.

Take

$$\boldsymbol{x}^* = \frac{d}{diag(\{L^1, L^{11}, L^3, L^4, L^5, L^6, L^7, L^8, L^{10}\})} =$$
$$= (0.1804, 0.0902, 0.0601, 0.0902, 0.0902, 0.0902, 0.0902, 0.1804, 0.1281)$$

Evaluate $J_1(\boldsymbol{x}^*, \boldsymbol{u}, \boldsymbol{t})$,

$$J_1(\boldsymbol{x}^*, \boldsymbol{u}, \boldsymbol{t}) = 0.0226.$$

$$J_1(\boldsymbol{x}^*, \boldsymbol{u}, \boldsymbol{t}) - d = 0.0027$$

It will consider after Iteration 2 the value of $J_1(\boldsymbol{x}^*, \boldsymbol{u}, \boldsymbol{t})$ is repeated means the value of $J_1^*(\boldsymbol{x}^*, \boldsymbol{u}, \boldsymbol{t})$ is 0.0226. Move the value of $J_1^*(\boldsymbol{x}^*, \boldsymbol{u}, \boldsymbol{t})$ into $\mathcal{J}^*$.

$$\mathcal{J}^* = \{0.0226\}.$$

According to above process, the CAM was used to find $J_2^*(\boldsymbol{x}^*, \boldsymbol{u}, \boldsymbol{t})$. All the process to find the optimum is similar to previous actions. For this example, CAM was used and after two iterations, it stopped; because the stopping criterion is satisfies and the minimizer is found:

$$\boldsymbol{u}^* = (0.1714, 0.0857, 0.0571, 0.0857, 0.0857, 0.0857, 0.0857, 0.1714, 0.0025)$$

$$J_2^*(\boldsymbol{x}, \boldsymbol{u}^*, \boldsymbol{t}) = 1$$

Move the value of $J_2^*(\boldsymbol{x}^*, \boldsymbol{u}, \boldsymbol{t})$ into $\mathcal{J}^*$. $\mathcal{J}^* = \{1, 0.0226\}$. The smallest element of $\mathcal{J}^*$ is the minimum of J (see Figures 6.1 and 6.2). Thus

$$J^* = 0.0226$$

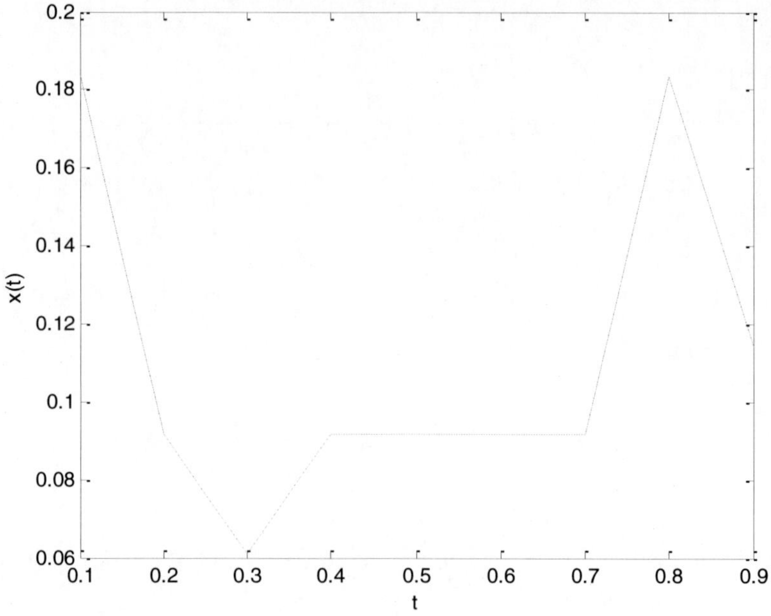

Figure 6.1. Optimum state vector of Example 6.1 for $n = 10$.

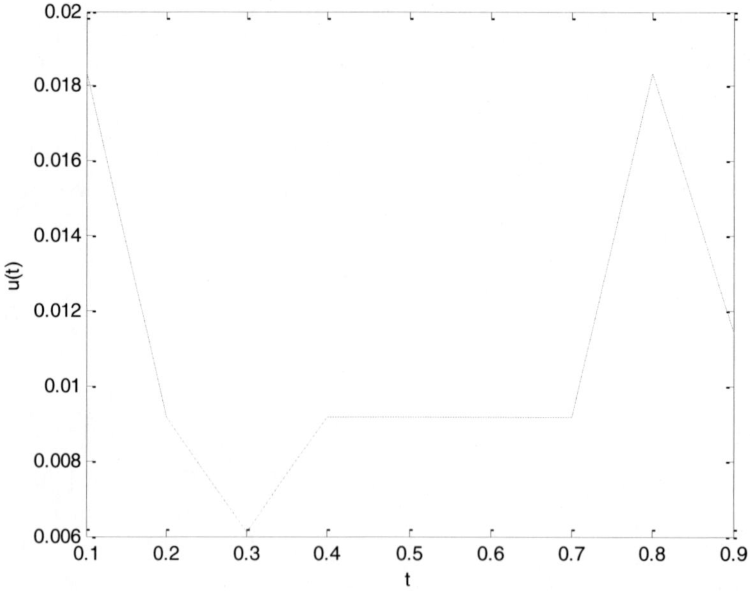

Figure 6.2. Optimum control vector of Example 6.1 for $n = 10$.

For $n = 20: t_1 = 0, t_2 = 0.05, \dots, t_{20} = 1$, the step size $h = 0.05$ and discretized OCP is

$$J = 0.05 \sum_{i=1}^{19} (\boldsymbol{x}^2(t) + \boldsymbol{u}(t))$$
$$s.t\ \boldsymbol{x} = 0.1 * \boldsymbol{u}(t)$$
$$\boldsymbol{x}(0) = 0, \boldsymbol{x}(1) = 0.5,$$

Firstly, the CAM is applied to find $\boldsymbol{x}^*$,

After two iterations, the criterion is satisfied and the following result was found,

$$\boldsymbol{x}^* = (0.1022, 0.0511, 0.0341, 0.0256, 0.0256, 0.0341, 0.0341, 0.0341, 0.0511, 0.0511, 0.0511, 0.0511, 0.0511, 0.0511, 0.0511, 0.0511, 0.0511, 0.1022, 0.0968)$$

and

$$J_1^*(\boldsymbol{x}^*, \boldsymbol{u}, t) = 0.0056$$

Similarly, we apply the CAM to find $\boldsymbol{u}^*$,

After two iterations, the criterion is satisfied and the following result was found,

$$\boldsymbol{u}^* = (0.1017, 0.0508, 0.0339, 0.0254, 0.0254, 0.0339, 0.0339, 0.0339, 0.0508, 0.0508, 0.0508, 0.0508, 0.0508, 0.0508, 0.0508, 0.0508, 0.0508, 0.1017, 0.0004)$$

and,

$$J_2^*(\boldsymbol{x}, \boldsymbol{u}^*, t) = 1.0935$$

Thus the global minimizer is the smallest elements of $\mathcal{J}^*$ (see Figures 6.3 and 6.4).

$$J^* = 0.0056.$$

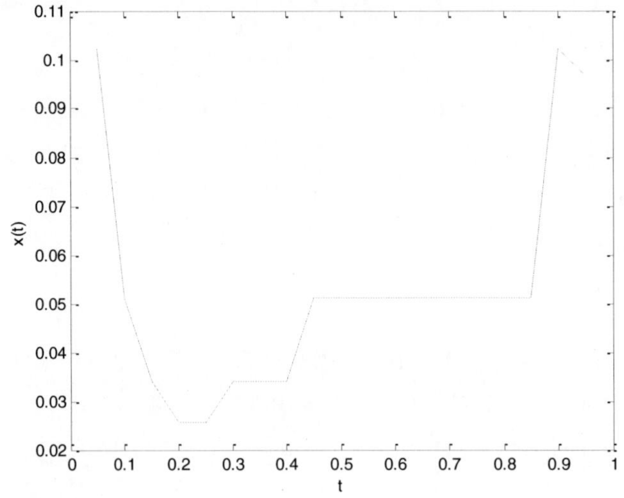

Figure 6.3. Optimum state vector of Example 6.1 for $n = 20$.

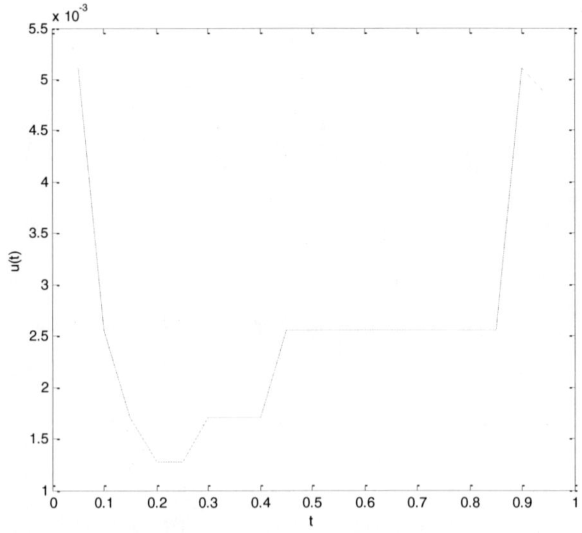

Figure 6.4. Optimum control vector of Example 6.1 for $n = 20$.

This example demonstrates that by increasing the appropriate number of partitions n the solution of the problem converges to global solution.

Example 6.2: The second minimization test functional (Rubio, 1985) is

$$J = \int_0^1 (\boldsymbol{x}_1^2(t) + \boldsymbol{x}_2^2(t))dt$$

$$s.t\, \dot{\boldsymbol{x}}_1 = \boldsymbol{x}_2(t), \dot{\boldsymbol{x}}_2 = 10\boldsymbol{x}_1^3(t) + \boldsymbol{u}(t)$$

$$\boldsymbol{x}_1(0) = 0, \boldsymbol{x}_1(1) = 0.1,$$

$$\boldsymbol{x}_2(0) = 0, \boldsymbol{x}_2(1) = 0.3.$$

The discretized form of above OCP is

$$J = \mathfrak{h} \sum_{i=1}^{n-1} (\boldsymbol{x}_1^2(t) + \boldsymbol{x}_2^2(t))$$

$$s.t\, \boldsymbol{x}_1 = \mathfrak{h} * \boldsymbol{x}_2(t)$$

$$\boldsymbol{x}_2 = \mathfrak{h} * 10\boldsymbol{x}_1^3(t) + \mathfrak{h} * \boldsymbol{u}(t)$$

$$\boldsymbol{x}_1(0) = 0, \boldsymbol{x}_1(1) = 0.1,$$

$$\boldsymbol{x}_2(0) = 0, \boldsymbol{x}_2(1) = 0.3.$$

For this problem, the interval $[0, 1]$ was partitioned into 10 equal subintervals. Thus, for $n = 10$: $t_1 = 0, t_2 = 0.1, \ldots, t_{10} = 1$. Then

CAM with three iterations was used to minimize the mentioned OCP and Criterion= 0.001.

The step size $h = 0.1$ and discretized OCP is

$$J = 0.1 \sum_{i=1}^{9} (x_1^2(t) + x_2^2(t))$$

$$s.t\, \boldsymbol{x}_1 = 0.1 * \boldsymbol{x}_2(t)$$

$$\boldsymbol{x}_2 = 0.1 * 10\boldsymbol{x}_1^3(t) + 0.1 * \boldsymbol{u}(t)$$

$$\boldsymbol{x}_1(0) = 0, \boldsymbol{x}_1(1) = 0.1,$$

$$\boldsymbol{x}_2(0) = 0, \boldsymbol{x}_2(1) = 0.3.$$

For this same example, the CAM is applied to find $\boldsymbol{x}_1{}^*$, $\boldsymbol{x}_2{}^*$ based on UVCT (see Tables 6.1 and 6.2 and Figures 6.5 and 6.6). The minimum value of the discretized OCP is 0.0218 by three iterations of CAM.

Table 6.1. Using CAM to find a local minimizer of Example 6.2 via UVCT (for $n = 10$)

iteration	X1	J1
k=1	(0.1998, 0.0666, 0.0499, 0.0666, 0.0666, 0.0666, 0.0999, 0.1998, 0.1842)	2.1624
k=2	(0.2023, 0.0674, 0.0506, 0.0674, 0.0674, 0.0674, 0.1011, 0.2023, 0.1740)	2.1713
k=3	(0.2044, 0.0681, 0.0511, 0.0681, 0.0681, 0.0681, 0.1022, 0.2044, 0.1654)	2.1808

Table 6.2. Using CAM for three iterations and finding a global minimizer of Example 6.2 via UVCT (for $n = 20$)

iteration	X2	J2
k=1	(0.1998, 0.0666, 0.0499, 0.0666, 0.0666, 0.0666, 0.0999, 0.1998, 0.1842)	0.0216
k=2	(0.2023, 0.0674, 0.0506, 0.0674, 0.0674, 0.0674, 0.1011, 0.2023, 0.1740)	0.0217
k=3	(0.2044, 0.0681, 0.0511, 0.0681, 0.0681, 0.0681, 0.1022, 0.2044, 0.1654)	0.0218

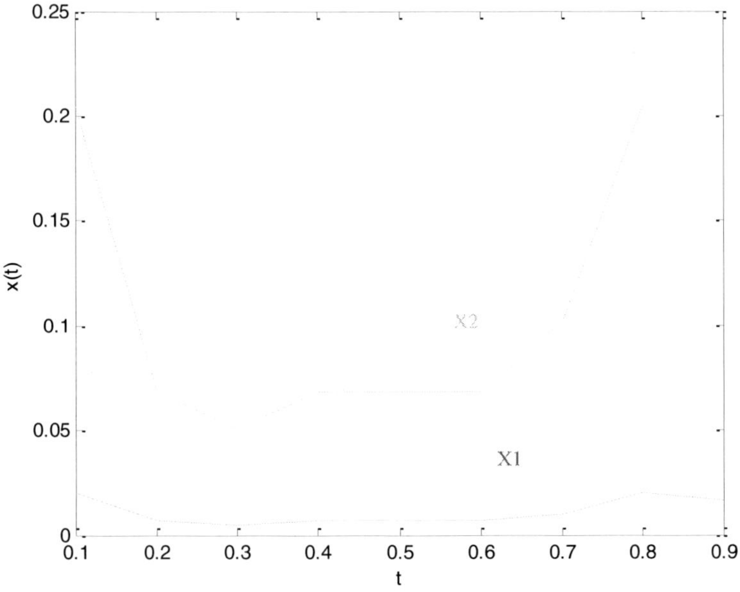

Figure 6.5. Optimum state vectors of Example 6.2 for $n = 10$.

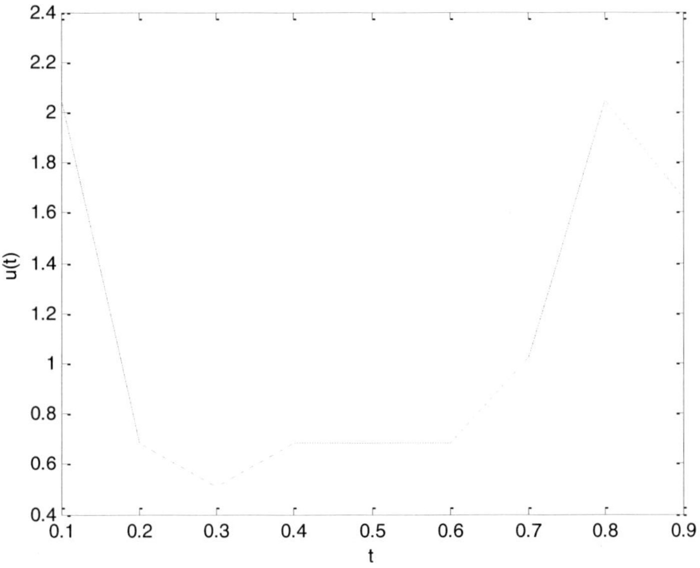

Figure 6.6. Optimum control vector of Example 6.2 for $n = 10$.

Example 6.3: Continuous Stirred Tank

The third minimization test functional is

$$J = \int_0^{0.78} (\boldsymbol{x}_1^2(t) + \boldsymbol{x}_2^2(t) + 0.1\boldsymbol{u}^2(t))dt$$

$$s.t\ \dot{\boldsymbol{x}}_1 = -(\boldsymbol{x}_1(t) + 0.25) + (\boldsymbol{x}_2(t) + 0.5)\exp\left(\frac{25\boldsymbol{x}_1(t)}{\boldsymbol{x}_1(t)+2}\right) - (1 + \boldsymbol{u}(t))(\boldsymbol{x}_1(t) + 0.25),$$

$$\dot{\boldsymbol{x}}_2 = 0.5 - \boldsymbol{x}_2(t) - (\boldsymbol{x}_2(t) + 0.5)\exp\left(\frac{25\boldsymbol{x}_1(t)}{\boldsymbol{x}_1(t) + 2}\right)$$

$$\boldsymbol{x}(0) = [0.05\ 0]^{\mathrm{T}}$$

The discretized form of above OCP is

$$J = \mathfrak{h}\sum_{i=1}^{n-1} (\boldsymbol{x}_1^2(t) + \boldsymbol{x}_2^2(t) + 0.1\boldsymbol{u}^2(t))$$

$$s.t\ \mathrm{x}_1 = \mathfrak{h} * (-(\boldsymbol{x}_1(t) + 0.25) + (\boldsymbol{x}_2(t) + 0.5)\exp\left(\frac{25\boldsymbol{x}_1(t)}{\boldsymbol{x}_1(t) + 2}\right) - (1 + \boldsymbol{u}(t))(\boldsymbol{x}_1(t) + 0.25))$$

$$\mathrm{x}_2 = \mathfrak{h} * (0.5 - \boldsymbol{x}_2(t) - (\boldsymbol{x}_2(t) + 0.5)\exp\left(\frac{25\boldsymbol{x}_1(t)}{\boldsymbol{x}_1(t) + 2}\right))$$

$$\boldsymbol{x}(0) = [0.05\ 0]^{\mathrm{T}}$$

For this problem, the interval $[0, 0.78]$ was partitioned into 10, 100, 200, 300 and 400 equal subintervals:

$$n = 10\colon \mathfrak{h}_1 = 0, \mathfrak{h}_2 = 0.1, \dots, \mathfrak{h}_{10} = 0.78;$$
$$n = 100\colon \mathfrak{h}_1 = 0, \mathfrak{h}_2 = 0.01, \dots, \mathfrak{h}_{10} = 0.78;$$
$$n = 200\colon \mathfrak{h}_1 = 0, \mathfrak{h}_2 = 0.005, \dots, \mathfrak{h}_{10} = 0.78;$$
$$n = 300\colon \mathfrak{h}_1 = 0, \mathfrak{h}_2 = 0.0033, \dots, \mathfrak{h}_{10} = 0.78;$$
$$n = 400\colon \mathfrak{h}_1 = 0, \mathfrak{h}_2 = 0.0026, \dots, \mathfrak{h}_{10} = 0.78.$$

where $\mathfrak{h} = \frac{0.78-0}{n}$ is step size of discretized OCP. Consequently, the step size $\mathfrak{h}$ is equal to $0.1, 0.01, 0.005, 0.0033, 0.0026$ respectively. Then CAM is used with stopping criterion$= 0.001$.

For $n = 10$:

$$\boldsymbol{x}_1^* = (0.2041, 0.1020, 0.0680, 0.1020, 0.1020, 0.1020, 0.1020, 0.2041, 0.0137)$$

$$\boldsymbol{x}_2^* = (-0.0113, -0.0057, -0.0038, -0.0057, -0.0057, -0.0057, -0.0057, -0.0113, -0.0008)$$
$$J_{n=10}^* = 0.3163$$

for $n = 100$, $n = 200$, $n = 300$, $n = 400$, the solutions by CAM are $J_{n=100}^* = 0.0937$, $J_{n=200}^* = 0.0583$, $J_{n=300}^* = 0.0402$, $J_{n=400}^* = 0.0290$, with step size $\mathfrak{h} = 0.01, 0.005, 0.0033, 0.0026$ respectively (see Figures 6.7, 6.8, 6.9, 6.10, 6.11, 6.12, 6.13, 6.14, 6.15 and 6.16). The minimum by three iterations of CAM is 0.0290. Additionally, this example was also solved by 10 iterations of DISOPE algorithm and the given result was 0.028. The result shows that the CAM converges to the global solution faster than DISOPE algorithm.

End of iteration 3.

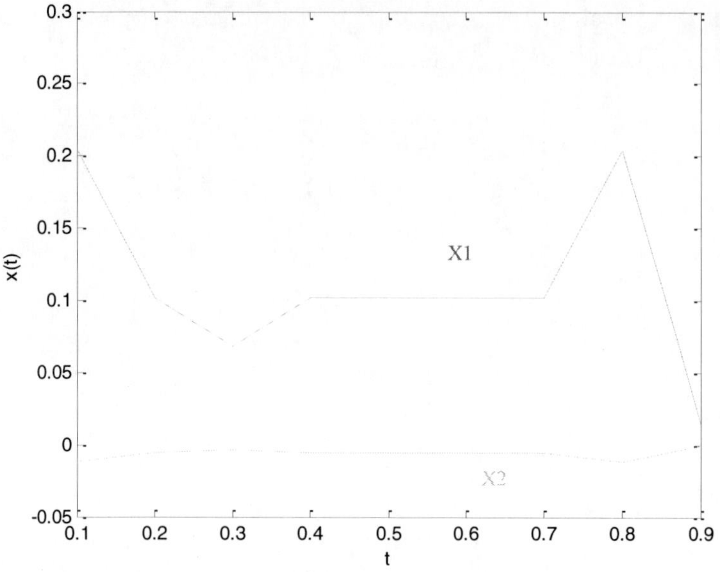

Figure 6.7. Optimum state vectors of Example 6.3 for $n = 10$.

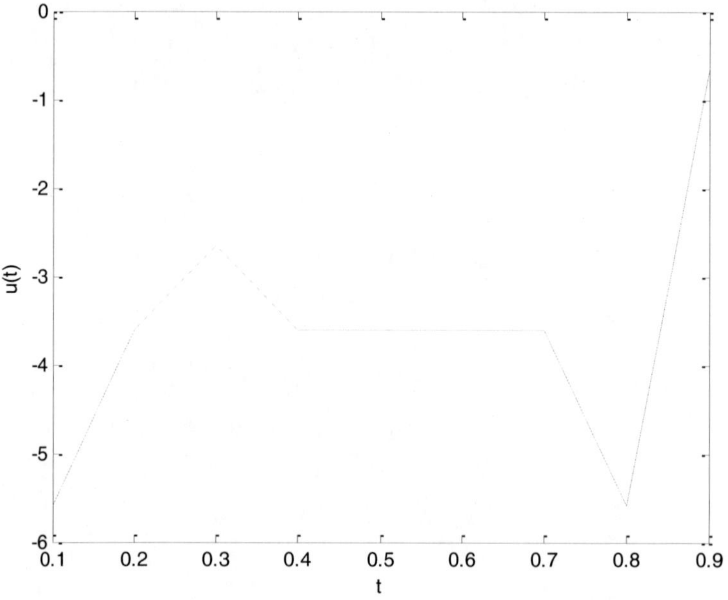

Figure 6.8. Optimum control vector of Example 6.3 for $n = 10$.

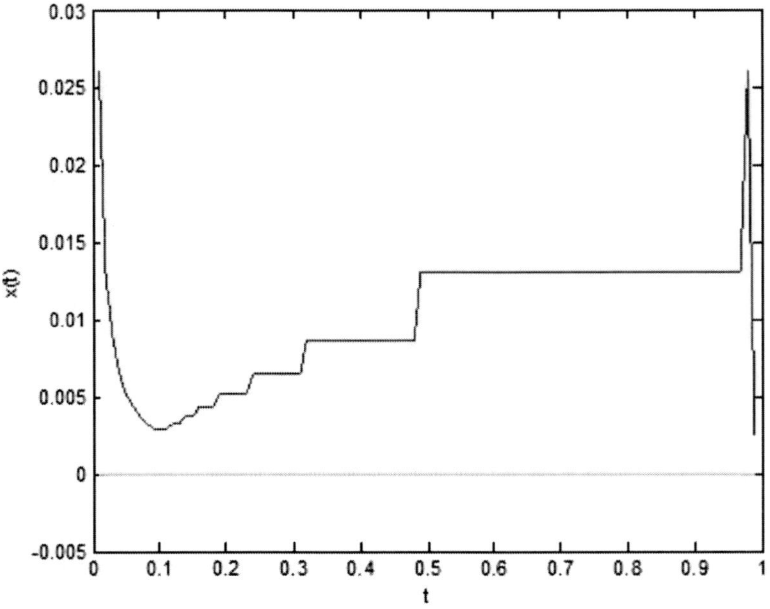

Figure 6.9. Optimum state vectors of Example 6.3 for $n = 100$.

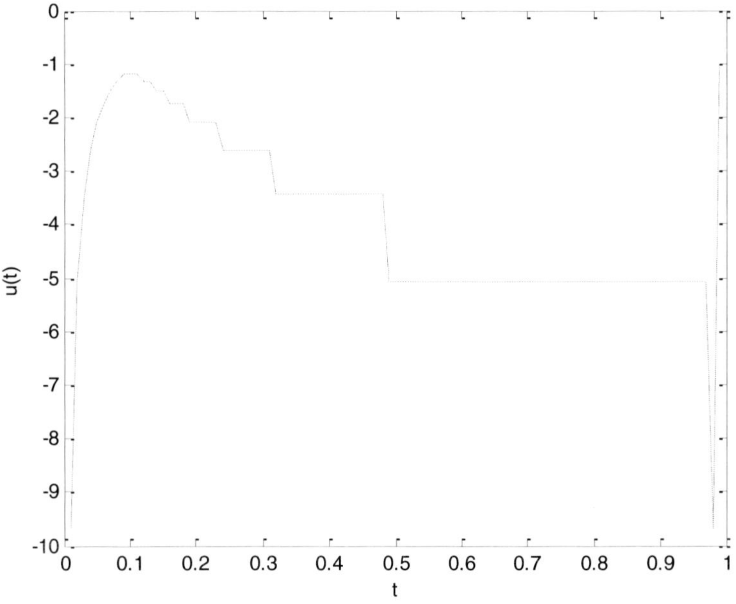

Figure 6.10. Optimum control vector of Example 6.3 for $n = 100$.

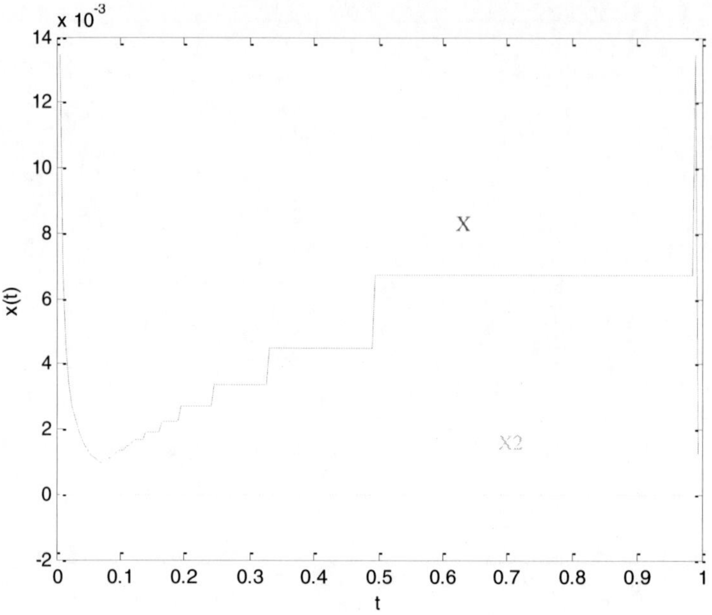

Figure 6.11. Optimum state vectors of Example 6.3 for $n = 200$.

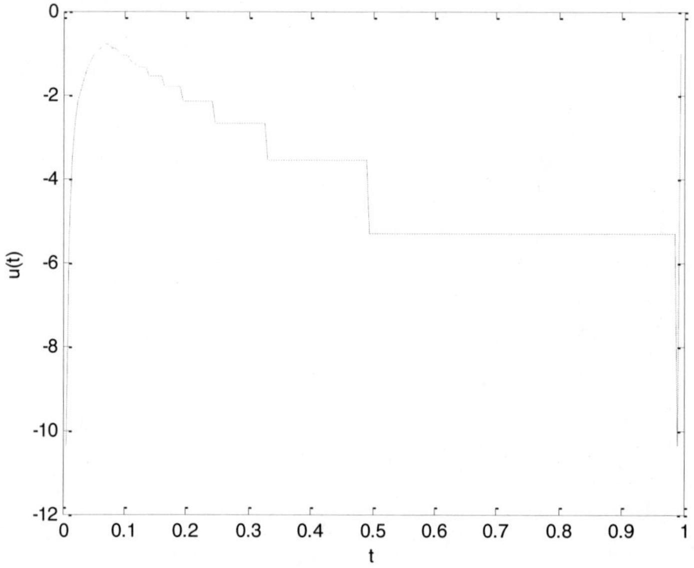

Figure 6.12. Optimum control vector of Example 6.3 for $n = 200$.

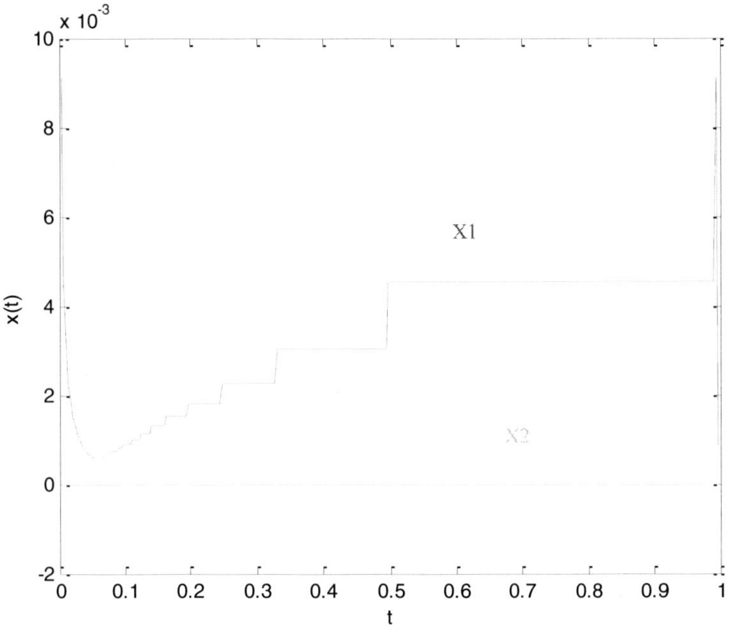

Figure 6.13. Optimum state vectors of Example 6.3 for $n = 300$.

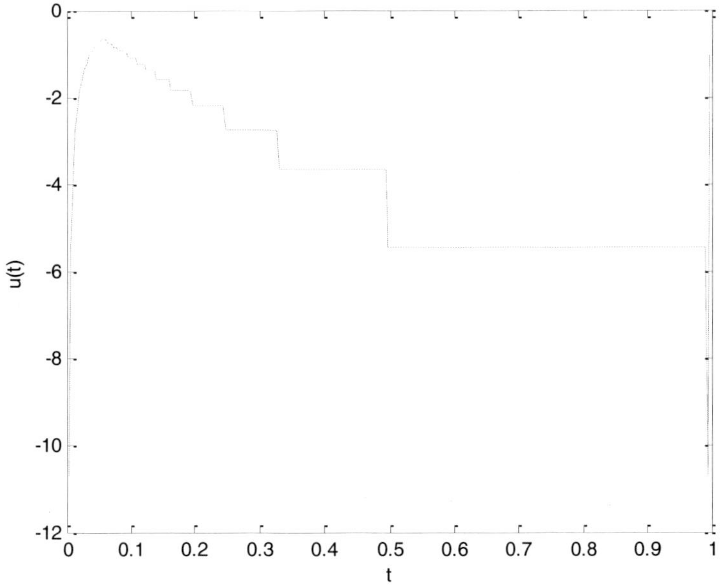

Figure 6.14. Optimum control vector of Example 6.3 for $n = 300$.

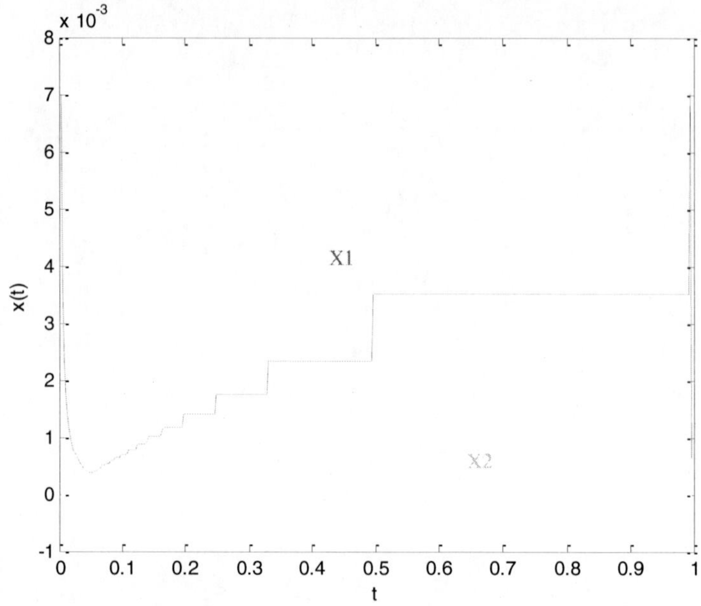

Figure 6.15. Optimum state vectors of Example 6.3 for $n = 400$.

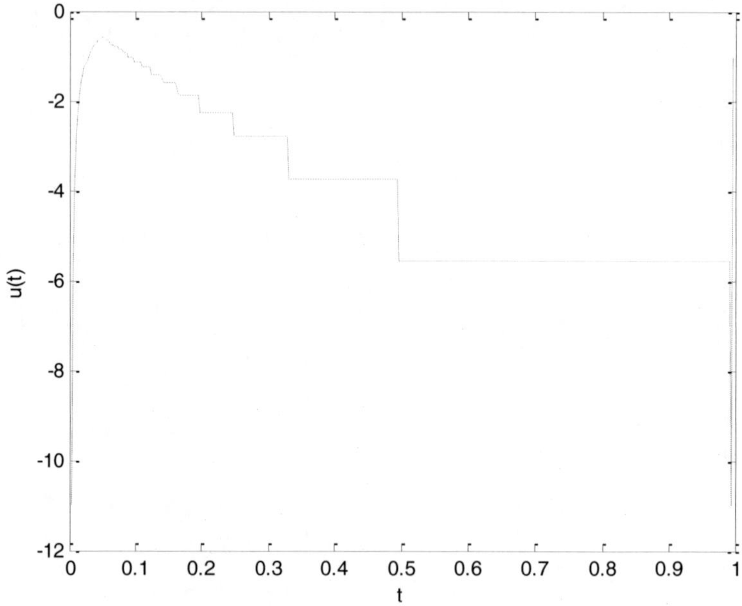

Figure 6.16. Optimum control vector of Example 6.3 for $n = 400$.

This numerical result implies that the CAM with a suitable number of partitions converges to the global solution and the main property of this method is it converges to global solutions in the lowest number of iterations.

Example 6.4: The forth minimization test functional [5] is

$$J = \int_0^1 (\boldsymbol{u}^2(t))dt$$

$$s.t\ \dot{\boldsymbol{x}}(t) = \boldsymbol{x}^2(t) + \boldsymbol{u}(t),$$

$$\boldsymbol{x}(0) = 0, \boldsymbol{x}(1) = 0.5,$$

The discretized form of above OCP is

$$J = \mathfrak{h} \sum_{i=1}^{n-1} (\boldsymbol{u}^2(t))$$

$$s.t\ \mathrm{x}_1 = \mathfrak{h} * \boldsymbol{x}^2(t) + \mathfrak{h} * \boldsymbol{u}(t)$$

$$\boldsymbol{x}(0) = 0, \boldsymbol{x}(1) = 0.5,$$

For this problem, the interval $[0, 1]$ was partitioned into 10 equal subintervals. Thus, for $n = 10: t_1 = 0, t_2 = 0.1, \dots, t_{10} = 1$. The CAM is applied to minimize the mentioned OCP with stop criterion= 0.01.

The step size $\hbar = 0.1$ and discretized OCP is

$$J = 0.1 \sum_{i=1}^{9} (\boldsymbol{u}^2(t))$$

$$s.t\, \boldsymbol{x}_1 = 0.1 * \boldsymbol{x}^2(t) + 0.1 * \boldsymbol{u}(t)$$

$$\boldsymbol{x}(0) = 0, \boldsymbol{x}(1) = 0.5,$$

For this example, the CAM is applied to find $\boldsymbol{x}^*$ and $\boldsymbol{u}^*$ based on UVCT.

$$\boldsymbol{x}^* = (0.0002, 0.0001, 0.0000, 0.0001, 0.0001, 0.0001, 0.0001, 0.0002, 0.0002)$$

$$\boldsymbol{u}^* = (0.1998, 0.0666, 0.0499, 0.0666, 0.0666, 0.0666, 0.0999, 0.1998, 0.1842)$$

The global solution of the discretized OCP is 0.0214 by two iterations of CAM (see Figures 6.17 and 6.18) and the result by using Genetic algorithm is 0.4447. This example was also solved by DISOPE algorithm and the result was 0.0000 in 12 iterations.

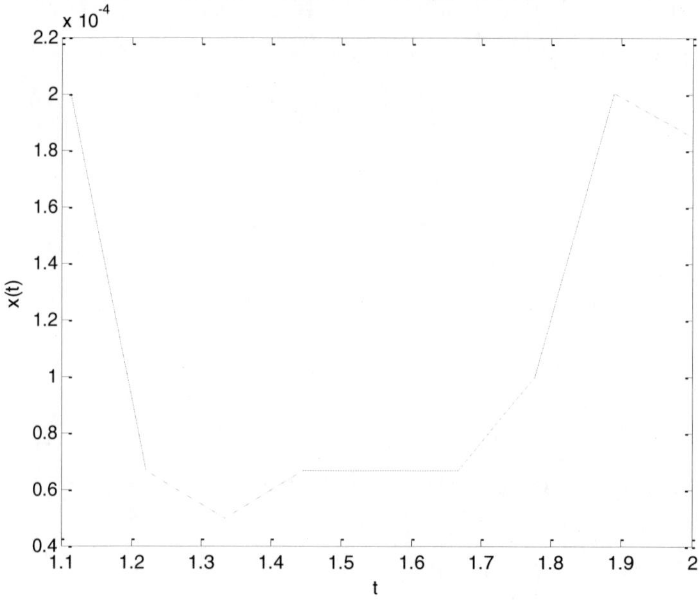

Figure 6.17. Optimum state vector of Example 6.4 for $n = 10$.

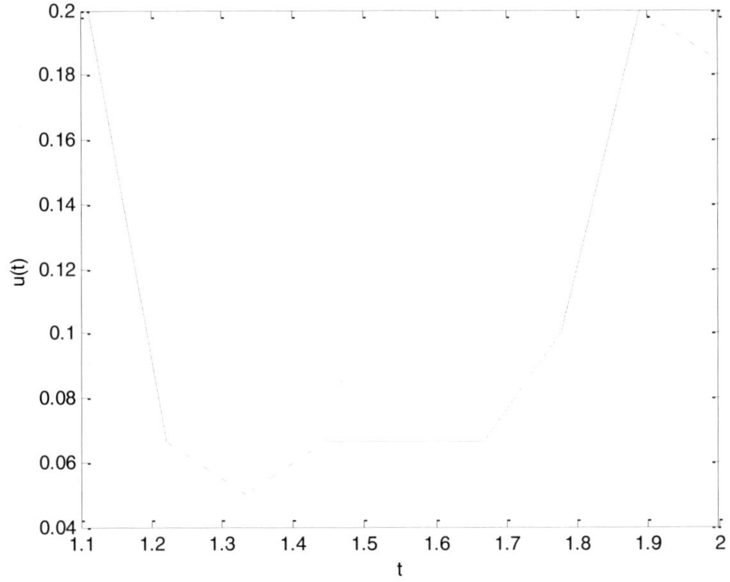

Figure 6.18. Optimum control vector of Example 6.4 for $n = 10$.

REFERENCES

[1] Singh, M. G. & Titli, A. (1978). *Systems: decomposition, optimisation, and control*: Pergamon.

[2] Becerra, V. M. (1994). *"Development and applications of novel optimal control algorithms,"* City University.

[3] Roberts, P. (1993), "An algorithm for optimal control of nonlinear systems with model-reality differences", in *Proceedings of 12th IFAC World Congress on Automatic Control*, 407-412.

[4] Leithold, L. (1971). *The Calculus Book.*

[5] Fard, O. S. & Borzabadi, A. H. (2007). "Optimal control problem, quasi-assignment problem and genetic algorithm", in *Proceedings of World Academy of Science, Engineering and Technology*, 70-43.

Chapter 7

A COMBINATION OF THE CUTTING ANGLE METHOD AND A LOCAL SEARCH ON OPTIMAL CONTROL PROBLEMS

ABSTRACT

A new technique is proposed to use the Cutting Angle Method (CAM) for solving optimal control problems. The study focuses on the combinations of the CAM with a local search for solving Optimal Control Problems (OCP). The numerical outcomes specified that our approach is very efficient to search the locations of the global optimums.

Keywords: optimal control problem, optimization, cutting angle method, dynamic integrated system optimization, and parameter estimation algorithm

7.1. INTRODUCTION

As mentioned before, Bagirov and Rubinov (2000) proposed the cutting angle method which is able to help local search algorithms to converge to global optimum. This method is utilized in combination with a local search algorithm to arrive at a truly global solution [1].

This section shall propose the CAM as the global search which will be able to help DISOPE algorithm as the local search to converge to global optimum. The result of this section shall confirm that the output of the DISOPE algorithm is global.

DISOPE is a technique for solving optimal control problems where there are differences in structure and parameter values between reality and the model employed in the computations. The main property of the procedure is that by iterating on appropriately modified model based problems the correct optimal solution is achieved in spite of the model-reality differences [2].

However, when solving realistic problems, the problem can often be well approximated by a linear-quadratic problem, that is a problem with linear (time-varying) dynamics and quadratic performance index whose (time-varying) weighting matrices are the second derivatives of the states and the controls [3].

An algorithm that uses this approach would not be solving the original problem but rather solves the simplified Linear Quadratic Regulator (LQR) problem. The solution to the LQR problem is hoped to approximate the real solution or in other words to converge to the real solution. Thus, a good algorithm has to take into account the model-reality differences to be successful. One such algorithm is the DISOPE algorithm. DISOPE discussed in this study uses LQR problems as models.

The combinations of the cutting angle method in global optimization and discrete gradient method (DGM) as a local search were presented by Bagirov and Rubinov (2003) which successfully confirmed their claim about CAM.

Particularly, DGM is a type of gradient descent method which CAM could help overcome with the above problems and help approach the exact solution by suitable choices of the support functions.

Based on the ability of CAM in global optimization and its previous combination with DG method, the combination of CAM and DISOPE algorithm is proposed as follows and is able to arrive at the global solution.

7.2. A Combination of CAM and DISOPE Algorithm for Solving the Continuous Time OCP

Based on the combination of CAM and a local search (DGM) by Bagirov and Rubinov (2003), from (6.3) assume that

$$J = \mathfrak{h} \sum_{i=1}^{n-1} E_i\,(\boldsymbol{x}, \boldsymbol{u}, t) = g(\boldsymbol{x}, \boldsymbol{u}, t)$$

Then the following problem of global optimization is studied,

$$g(\boldsymbol{x}, \boldsymbol{u}, t) \rightarrow \text{min subject to } (\boldsymbol{x}, \boldsymbol{u}, t) \in \mathcal{S} \tag{7.1}$$

where

$$\mathcal{S} = \{\boldsymbol{x}(t) = (\boldsymbol{x}_1, \dots, \boldsymbol{x}_m) \in \mathbb{R}^n : \sum\nolimits_1^m \boldsymbol{x}_i = 1\,, \boldsymbol{x}_1 \geq 0, \dots, \boldsymbol{x}_m \geq 0;$$

$$\boldsymbol{u}(t) = (\boldsymbol{u}_1, \dots, \boldsymbol{u}_m) \in \mathbb{R}^n : \sum\nolimits_1^m \boldsymbol{u}_i = 1\,, \boldsymbol{u}_1 \geq 0, \dots, \boldsymbol{u}_m \geq 0\}$$

is the unit simplex and g is a ICAR function defined on $\mathcal{S}$ (unit simplex).

Applying the CAM to the minimization of the function g is not able to take into account the known value of this function at stationary vector $\mathcal{y}$ which was found by DISOPE algorithm as a local search. In order to use this value, a function g shall be transformed into a new function p.

A function p is called a transformed function of g (with respect to vector y) if

1. $p(\boldsymbol{x},\boldsymbol{u},t) \leq g(\boldsymbol{x},\boldsymbol{u},t)$;
2. $$\min_{x,u\in\mathcal{S}} p(\boldsymbol{x},\boldsymbol{u},t) = \min_{x,u\in\mathcal{S}} g(\boldsymbol{x},\boldsymbol{u},t). \quad (7.2)$$

It follows from (7.2) that the set $\mathcal{T}(\boldsymbol{x}^*,\boldsymbol{u}^*,t) = \{\boldsymbol{x},\boldsymbol{u} \in \mathcal{S}: g(\boldsymbol{x},\boldsymbol{u},t) \leq p(\boldsymbol{x}^*,\boldsymbol{u}^*,t)\}$ is nonempty for all $\boldsymbol{x}^*,\boldsymbol{u}^*$. Indeed if $(\overline{\boldsymbol{x}^*},\overline{\boldsymbol{u}^*},t)$ is a global minimizer of g over $\mathcal{S}$ then $g(\overline{\boldsymbol{x}^*},\overline{\boldsymbol{u}^*},t) = \min_{x,u\in\mathcal{S}} p(\boldsymbol{x},\boldsymbol{u},t) \leq p(\boldsymbol{x}^*,\boldsymbol{u}^*,t)$ so $(\overline{\boldsymbol{x}^*},\overline{\boldsymbol{u}^*},t) \in \mathcal{T}(\boldsymbol{x}^*,\boldsymbol{u}^*,t)$.

Proposition 7.1: The following functions are transformed for a function g with respect to $(\boldsymbol{y},\boldsymbol{u},t)$, [4]:

1. $p_1(\boldsymbol{x},\boldsymbol{u},t) = \min(g(\boldsymbol{x},\boldsymbol{u},t), g(\boldsymbol{y},\boldsymbol{u},t))$;
2. $p_2(\boldsymbol{x},\boldsymbol{u},t) = \min_{j=1,..,n} \min_{b\in\mathcal{A}_j} p_1(b(\boldsymbol{x},\boldsymbol{u},t) + (1-b)(\boldsymbol{x},\boldsymbol{u},t)^j)$

   where $1 \in \mathcal{A}_j \subset [0,1]$, $(\boldsymbol{x},\boldsymbol{u},t)^j \in \mathcal{S}, j = 1,..,n$;
3. $p_3(\boldsymbol{x},\boldsymbol{u},t) = \min_{j=1,..,n} \min_{b\in\mathcal{A}_j} g(b(\boldsymbol{x},\boldsymbol{u},t) + (1-b)(\boldsymbol{x},\boldsymbol{u},t)^j)$

   where $\mathcal{A}$ and $(\boldsymbol{x},\boldsymbol{u},t)^j$ are as in item 2);
4. $p_4(\boldsymbol{x},\boldsymbol{u},t) = \min_{b\in B} \min_{a\in\mathcal{A}} \sum_{j\in\mathcal{J}} b_j\, g(b(\boldsymbol{x},\boldsymbol{u},t) + (1 - b)(\boldsymbol{x},\boldsymbol{u},t)^j)$, where $\mathcal{J}$ is a finite set of indices,

$$B = \left\{ (b_j)_{j\in\mathcal{J}} : \sum_{j\in\mathcal{J}} b_j = 1, b_j \geq 0 \ (j \in \mathcal{J}) \right\},$$

$$1 \in \mathcal{A}_j \subset [0,1],\ (\boldsymbol{x},\boldsymbol{u},t)^j \in \mathcal{S}\ (j \in \mathcal{J});$$

5. $p_5(\boldsymbol{x}, \boldsymbol{u}, t) = \min_{b \in \mathcal{B}} \min_{a \in \mathcal{A}} \sum_{j \in \mathcal{J}} b_j\, p_1(\boldsymbol{x}, \boldsymbol{u}, t)(b(\boldsymbol{x}, \boldsymbol{u}, t) + (1 - b)(\boldsymbol{x}, \boldsymbol{u}, t)^{j})$, where $\mathcal{J}, \mathcal{B}, \mathcal{A}$ are as in item 4);
6. $p_6(\boldsymbol{x}, \boldsymbol{u}, t) = b p_1(\boldsymbol{x}, \boldsymbol{u}, t) + (1 - b) g(\boldsymbol{x}, \boldsymbol{u}, t)$ where $b \in (0,1)$.

Furthermore, the choice of the sets $\mathcal{A}_j$, $j = 1, .., n$, depends on the problem under consideration and in particular, on the number of vectors. The number of elements of $\mathcal{A}_j$ should be large enough in order to obtain a good minorant for the objective function g. On the other hand it should not be too large, otherwise a large number of objective function evaluations at each iteration of the cutting angle method shall appear.

Remark 7.1:

(a) functions $p_2 - p_5$ depend on many parameters.
(b) All functions $p_1 - p_6$ are nonsmooth [4].

Consider the problem of global optimization presented by (7.1). An algorithm is proposed for its solution. This algorithm is based on a combination of the CAM and a local search and a choice of transformed function p which was proposed by Bagirov and Rubinov (2003).

The Algorithm (CAM-DISOPE)

Step 0. Initialization

Choose an arbitrary starting vectors $\overline{(\boldsymbol{x}, \boldsymbol{u}, t)}^0$. Set $i = 0$.

Step 1. Find stationary vectors of g over S by the DISOPE algorithm, starting from the vectors $(\overline{\boldsymbol{x}}, \overline{\boldsymbol{u}}, t)^i$. Denote this stationary vector by $(\overline{\boldsymbol{x}}, \overline{\boldsymbol{u}}, t)^i$ and let $g_i = g((\overline{\boldsymbol{x}}, \overline{\boldsymbol{u}}, t)^i)$.

Step 2. Construct a transformed function p_i of the function g with respect to the vector $(\boldsymbol{x}, \boldsymbol{u}, t)^i$.

Step 3. Take vectors $(\boldsymbol{x}, \boldsymbol{u}, t)^r = e^r$, $r = 1,..,n$, $(\boldsymbol{x}, \boldsymbol{u}, t)^{n+1} = (\boldsymbol{x}, \boldsymbol{u}, t)^i$. Let $l^r = p^i((\boldsymbol{x}, \boldsymbol{u}, t)^r)/(\boldsymbol{x}, \boldsymbol{u}, t)^r$, $r = 1,..,n+1$ and set $j = n+1$. Construct the function $h_j\big((\boldsymbol{x}, \boldsymbol{u}, t)\big) = \max_{r \le j} \min_{i \in \mathcal{I}(l^r)} l_i^r (\boldsymbol{x}, \boldsymbol{u}, t)_i$, $l^r = (l_1^r, \ldots, l_n^r) \in \mathbb{R}^n, l_i^r \ge 0, i = 1, \ldots, n, \mathcal{I}(l^r) = \{i: l_i^r > 0\}, r = 1, \ldots, j, j \ge n$.

Step 4. Solve the problem

$$h_j\big((\boldsymbol{x}, \boldsymbol{u}, t)\big) \to \min \text{ subject to } (\boldsymbol{x}, \boldsymbol{u}, t) \in \mathcal{S}. \tag{7.3}$$

Step 5. Let $(\mathbf{x}^*, \boldsymbol{u}^*, t)$ be a solution of the problem (7.3). Set $j = j + 1$ and $(\boldsymbol{x}, \boldsymbol{u}, t)^j = (\mathbf{x}^*, \boldsymbol{u}^*, t)$.

Step 6. Compute $p^* = p^i((\mathbf{x}^*, \boldsymbol{u}^*, t))$. If $p^* < g_i$ then find a vector $(\boldsymbol{x}, \boldsymbol{u}, t)' \in \mathcal{T}((\mathbf{x}^*, \boldsymbol{u}^*, t))$, set $i = i + 1$, $(\overline{\boldsymbol{x}}, \overline{\mathbf{u}}, t)^i = (\boldsymbol{x}, \boldsymbol{u}, t)'$ and go Step 1.

Step 7. Otherwise compute $l^j = p^i((\boldsymbol{x}, \boldsymbol{u}, t)^j)/(\boldsymbol{x}, \boldsymbol{u}, t)^j$, define the function

$$h_j(\boldsymbol{x}, \boldsymbol{u}, t) \to \max\{h_{j-1}(\boldsymbol{x}, \boldsymbol{u}, t) \to, \min_{i \in \mathcal{I}(l^j)} l_i^j (\boldsymbol{x}, \boldsymbol{u}, t_i)\}$$
$$\equiv \max_{r \le j} \min_{i \in \mathcal{I}(l^r)} l_i^r (\boldsymbol{x}, \boldsymbol{u}, t)_i$$

Check the stopping criterion, if it is satisfied stop, else go Step 4.

The validation of the globality of DISOPE algorithm used as a local search for CAM will be based on the assumption of the following proposition.

Proposition 7.2: Assume that the function g has a finite number of stationary vectors. Then the CAM algorithm terminates after a finite number of iterations at a global minimizer of g [4].

A major property of the combination of CAM and DISOPE algorithm is that the combination converges to global optimum based on the Proposition 7.2. Furthermore, the ability of CAM is it can be combined with a local search and it will converge to a global solution. But if the output of the local search was global in the first place, the output of the combination with CAM will be the same value and that confirms the globality. Thus, it implies that DISOPE algorithm converges to the global solution via LQR model.

The following example was solved by the above algorithm to show that the globality of the combination of CAM and DISOPE algorithm is valid.

7.3. Some Numerical Examples Solved by CAM-DISOPE Algorithm Using UVCT

This subsection resolves Example 6.3 by the combination algorithm to demonstrate that the globality of DISOPE is valid.

Example 7.1: Resolve the Example 7.3 of Chapter 6 by above method

$$J = \int_0^{0.78} (\boldsymbol{x}_1^2(t) + \boldsymbol{x}_2^2(t) + 0.1\boldsymbol{u}^2(t))dt$$

s.t

$$\dot{\boldsymbol{x}}_1 = -(\boldsymbol{x}_1(t) + 0.25) + (\boldsymbol{x}_2(t) + 0.5)\exp\left(\frac{25\boldsymbol{x}_1(t)}{\boldsymbol{x}_1(t)+2}\right) - (1 + \boldsymbol{u}(t))(\boldsymbol{x}_1(t) + 0.25),$$

$$\dot{\boldsymbol{x}}_2 = 0.5 - \boldsymbol{x}_2(t) - (\boldsymbol{x}_2(t) + 0.5)\exp\left(\frac{25\boldsymbol{x}_1(t)}{\boldsymbol{x}_1(t) + 2}\right)$$

$$\boldsymbol{x}(0) = [0.05\ 0]^{\mathrm{T}}$$

The discretized form of above OCP is

$$J = 0.0125 \sum_{i=1}^{79} (\boldsymbol{x}_1^2(t) + \boldsymbol{x}_2^2(t) \ + 0.1\boldsymbol{u}^2(t))$$

s.t

$$\boldsymbol{x}_1 = 0.125 * (-(\boldsymbol{x}_1(t) + 0.25) + (\boldsymbol{x}_2(t) + 0.5) \exp\left(\frac{25\boldsymbol{x}_1(t)}{\boldsymbol{x}_1(t) + 2}\right) - (1 + \boldsymbol{u}(t))(\boldsymbol{x}_1(t) + 0.25))$$

$$\boldsymbol{x}_2 = 0.125 * (0.5 - \boldsymbol{x}_2(t) - (\boldsymbol{x}_2(t) + 0.5) \exp\left(\frac{25\boldsymbol{x}_1(t)}{\boldsymbol{x}_1(t) + 2}\right))$$

$$\boldsymbol{x}(0) = [0.05\ 0]^{\mathrm{T}}$$

For this problem, the interval $[0, 0.78]$ was partitioned into 80 equal subintervals.

On the other hand, the result of applying the DISOPE algorithm is $= 0.1041$.

Utilizing CAM to find $\boldsymbol{x}_1^*$ for $n = 80$:

Step 0. Initialization

Take vectors $e^1 = (1,0,0,\dots,0)_{1\times79}, e^2 = (0,1,0,\dots,0)_{1\times79}, \dots, e^{79} = (0,\dots,0,1)_{1\times79}$ and construct basis vectors,

Step 1.

Find stationary vectors of discretized, $(\boldsymbol{x}_1\text{local}, \boldsymbol{x}_2\text{local}, \boldsymbol{u}\text{local}, t)$, by the DISOPE algorithm and starting from the vectors.

$$\boldsymbol{x}\text{local} = (\boldsymbol{x}_1\text{local};\ \boldsymbol{x}_2\text{local}) =$$

$$(0.05, 0.0484, 0.0469, \dots, 0.0292, 0.0299;$$
$$0., -0.0041, -0.0079, \dots, -0.0658, -0.0664)_{2\times 79}$$

$$\boldsymbol{u}\text{local} = (0.05, 0.0484, 0.0469, \dots, 0.0292, 0.0299)_{1\times 79}$$

$$J\text{local}(\boldsymbol{x}\text{local}, \boldsymbol{u}\text{local}, t) = 0.1041$$

Step 2. Compute

$$J_1^1(e^1, \boldsymbol{x}_2, \boldsymbol{u}, t) = 5.4525, J_1^2(e^2, \boldsymbol{x}_2, \boldsymbol{u}, t) = 10.9050, \dots,$$

$$J_1^{79}(e^1, \boldsymbol{x}_2, \boldsymbol{u}, t) = 5.4525$$

Step 3. Construct a transformed function L based on Proposition 6.1

$$J^1 = \min(J_1^1(e^1, \boldsymbol{x}_2, \boldsymbol{u}, t), J\text{local}) = 0.1041$$

$$\varsigma = 2.20204 \times 10^{-16}$$

$$L^1 = \left(\frac{(J^1 \times \varsigma) + (1-\varsigma) \times J_1^1(e^1, \text{x}_2, \text{u}, t)}{e^1}\right) = (5.4525, \infty, \infty, \dots, \infty)_{1\times 79}$$

$$J^2 = \min(J_1^2(e^2, \boldsymbol{x}_2, \boldsymbol{u}, t), J\text{local}) = 0.1041$$

$$L^2 = \left(\frac{(J^2 \times \varsigma) + (1-\varsigma) \times J_1^2(e^2, \boldsymbol{x}_2, \boldsymbol{u}, t)}{e^2}\right) = (\infty, 10.9050, \infty, \dots, \infty)_{1\times 79}$$

.

.

$$J^{79} = \min(\text{J}_1^{79}(e^{79}, \text{x}_2, \text{u}, t), J\text{local}) = 0.1041$$

$$L^{79} = \left(\frac{(J^{79} \times \varsigma) + (1-\varsigma) \times J_1^{79}(e^{79}, \boldsymbol{x}_2, \boldsymbol{u}, t)}{e^{79}}\right) = (\infty, \infty, \dots, \infty, 5.4525)_{1\times 79}$$

Step 4. Compute L^{80} for $\boldsymbol{x}_1$local

$$J^{80} = \min(J_1^{80}(\boldsymbol{x}_1\text{local}, \boldsymbol{x}_2, \boldsymbol{u}, t), J\text{local}) = 0.1041$$

$$L^{80} = \left(\frac{(J^{80}\times \mathfrak{s})+(1-\mathfrak{s})\times J_1^{80}(x_1\text{local}, \boldsymbol{x}_2, \boldsymbol{u}, t)}{\text{x}_1\text{local}}\right) = (109.0502, 112.6552, \ldots, 186.7298, 182.3582)_{1\times 79}$$

Step 5. Find $\boldsymbol{x}_1^*$ by CAM and compute $J_1^*(\boldsymbol{x}_1^*, \boldsymbol{x}_2, \boldsymbol{u}, t)$

$$\boldsymbol{x}_1^* = (0.0323, 0.0161, \ldots, 0.323, 0.0010)_{1\times 79}$$

$$J_1^*(\boldsymbol{x}_1^*, \boldsymbol{x}_2, \boldsymbol{u}, t) = 0.1111$$

Similarly, we used CAM to find $\boldsymbol{x}_2^*$, $\boldsymbol{u}^*$ and the smallest element J^* via UVCT (see Figures 7.1 and 7.2).

Finally, the algorithm obtains the following solution.

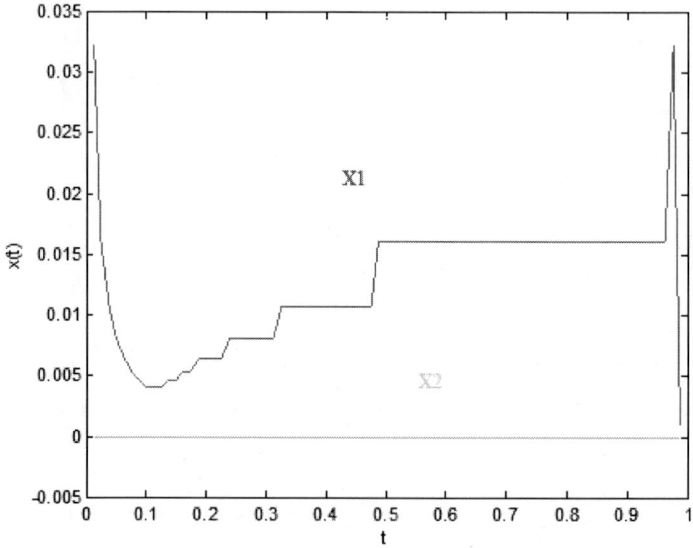

Figure 7.1. Optimum state vectors of Example 7.1 for $n = 80$.

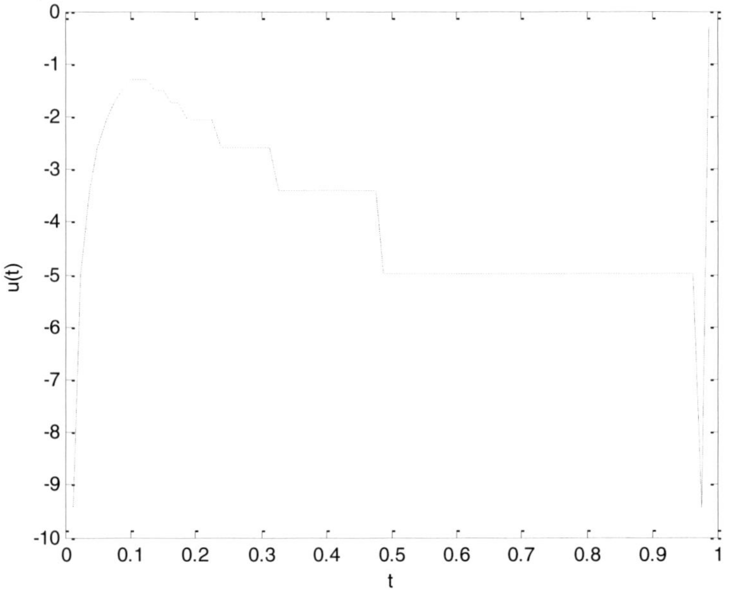

Figure 7.2. Optimum control vector of Example 7.1 for $n = 80$.

$$\boldsymbol{x}_1{}^* = (0.0323, 0.0161, \ldots, 0.323, 0.0010)_{1\times 79}$$

$$\boldsymbol{x}_2{}^* = (0.0323, 0.0161, \ldots, 0.323, 0.0010)_{1\times 79}$$

$$\boldsymbol{u}^* = (-9.4278, -4.9996, \ldots, -9.4278, -0.3170)_{1\times 79}$$

$$J^* = 0.1111$$

Example 7.2: Consider the following OCP

$$\text{minimize } J = \int_0^1 \boldsymbol{u}^2(t)dt$$

s.t

$$\dot{\boldsymbol{x}}(t) = \boldsymbol{x}^2(t) + \boldsymbol{u}(t)$$

$$\boldsymbol{x}(0) = 0, \boldsymbol{x}(1) = 0.5$$

The discretized form of above OCP is

$$J = 0.0099 \sum_{i=1}^{102-1} \boldsymbol{u}^2(t)$$

s.t

$$\boldsymbol{x}(t) = 0.0099 * (\boldsymbol{x}^2(t) + \boldsymbol{u}(t))$$

$$\boldsymbol{x}(0) = 0, \boldsymbol{x}(1) = 0.5$$

where $n = 102$ and step size $\mathfrak{h} = 0.0099$. For this problem, the interval $[0, 1]$ was partitioned into 102 equal subintervals. On the other hand, the result of this example with DISOPE algorithm is $J^* = 0.000055$ at $\boldsymbol{x} = 0.08245$. The combination of CAM and DISOPE algorithm were also used to solve this example via UVCT. The global solution of the discretized OCP is $J^* = 0.000055$ at

$$\boldsymbol{x} = (0.09, 0.090045164478881, 0.090081894249267, \ldots, 0.082453491544067)_{1\times 101}$$

with two iterations of the CAM (see Figures 7.3 and 7.4).

Example 7.3: Consider the following OCP

$$\text{minimize } J = \int_0^1 \boldsymbol{u}^2(t)dt$$

s.t

$$\dot{\boldsymbol{x}}(t) = \frac{1}{2}\boldsymbol{x}^2(t)\sin(\boldsymbol{x}(t)) + \boldsymbol{u}(t)$$

$$\boldsymbol{x}(0) = 0, \boldsymbol{x}(1) = 0.5$$

The discretized form of above OCP is

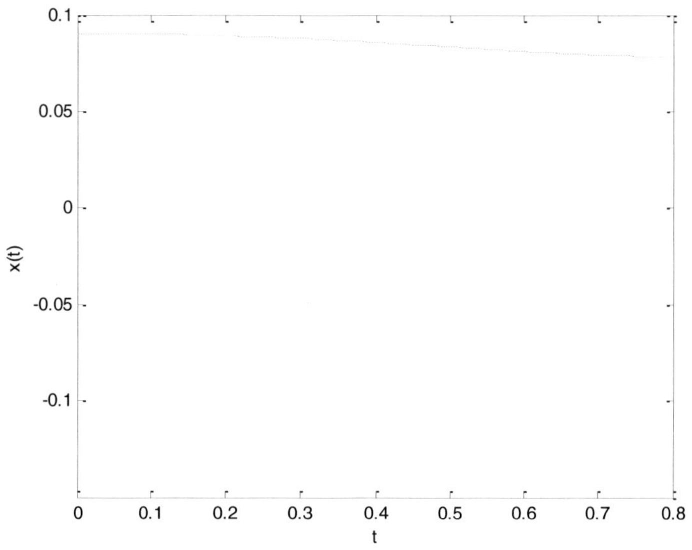

Figure 7.3. Optimum state vectors of Example 7.2 for $n = 102$.

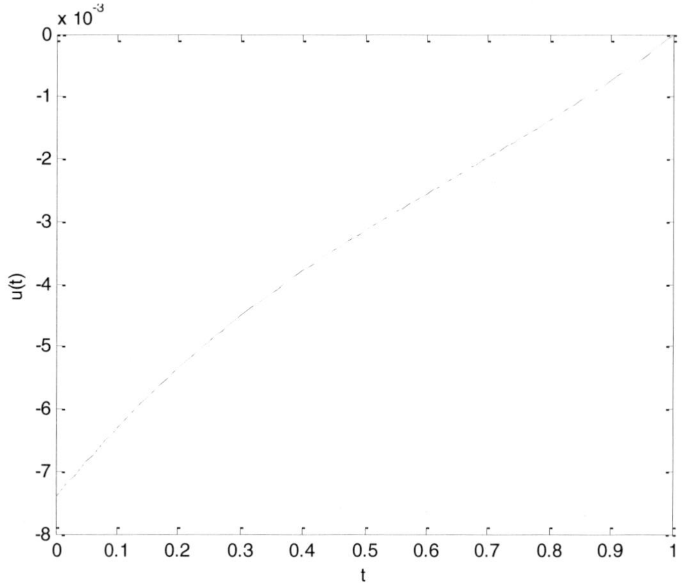

Figure 7.4. Optimum control vectors of Example 7.2 for $n = 102$.

$$J = 0.0099 \sum_{i=1}^{102-1} \boldsymbol{u}^2(t)$$

s.t

$$\boldsymbol{x}(t) = 0.0099 * (\frac{1}{2}\boldsymbol{x}^2(t)\sin(\boldsymbol{x}(t)) + \boldsymbol{u}(t))$$

$$\boldsymbol{x}(0) = 0, \boldsymbol{x}(1) = 0.5$$

where $n = 102$ and step size $\mathfrak{h} = 0.0099$. For this problem, the interval $[0, 1]$ was partitioned into 102 equal subintervals. On the other hand, the result of this example with DISOPE algorithm is $J^* = 0.0000005$. The CAM were used to solve this example via UVCT. The global minimum of the discretized OCP is $J^* = 0.0000005$ at

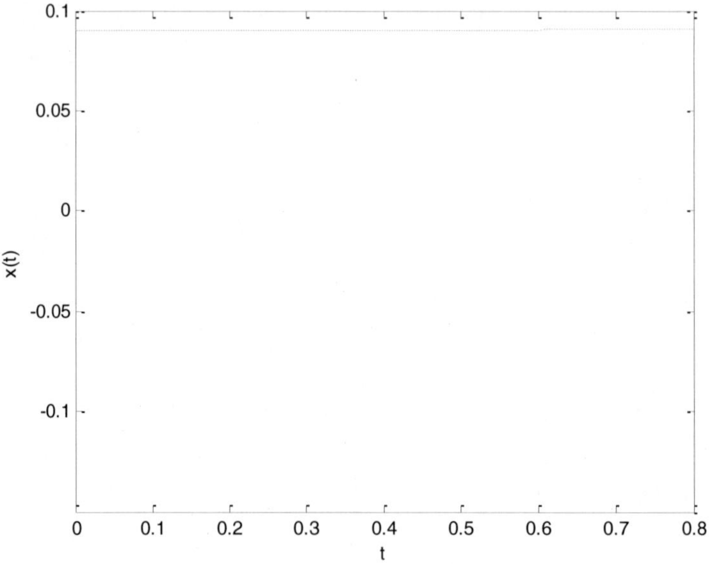

Figure 7.5. Optimum state vectors of Example 7.3 for $n = 102$.

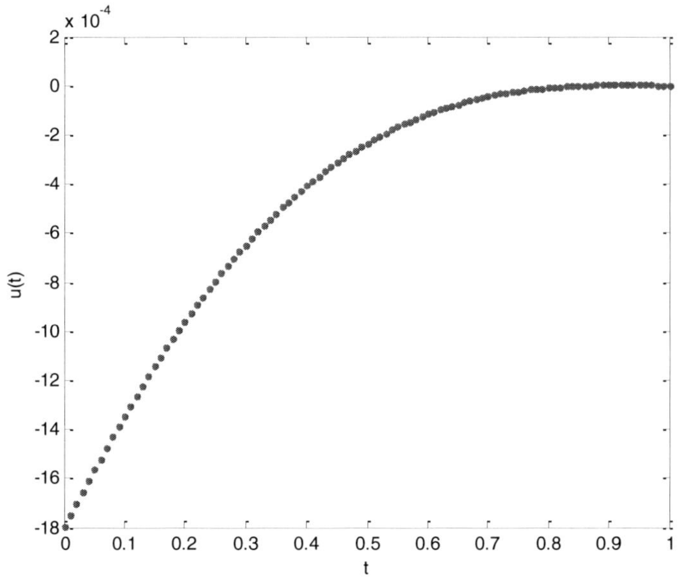

Figure 7.6. Optimum control vectors of Example 7.3 for $n = 102$.

$$\boldsymbol{x} = (0.09, 0.090004070867621, 0.090008133407132, \dots, 0.091426903511376)_{1\times 101}$$

with two iterations of CAM (see Figures 7.5 and 7.6).

Example 7.4: Consider the following OCP

$$\text{minimize } J = \int_0^1 (\boldsymbol{x}^3(t) + \boldsymbol{u}^2(t))dt$$

s.t

$$\dot{x}(t) = \boldsymbol{u}(t)$$

$$\boldsymbol{x}(0) = 0, \boldsymbol{x}(1) = 0.5$$

The discretized form of above OCP is

$$J = 0.0099 \sum_{i=1}^{102-1} (\boldsymbol{x}^3(t) + \boldsymbol{u}^2(t))$$

s.t

$$\boldsymbol{x}(t) = 0.0099 * \boldsymbol{u}(t)$$

$$\boldsymbol{x}(0) = 0, \boldsymbol{x}(1) = 0.5$$

where $n = 102$ and step size $\mathfrak{h} = 0.0099$. For this problem, the interval $[0, 1]$ was partitioned into 102 equal subintervals. The result of this example with DISOPE algorithm is $J^* = 25.04115$. The CAM here were used to solve this example via UVCT. The global solution of the discretized OCP is $J^* = 25.04115$ at

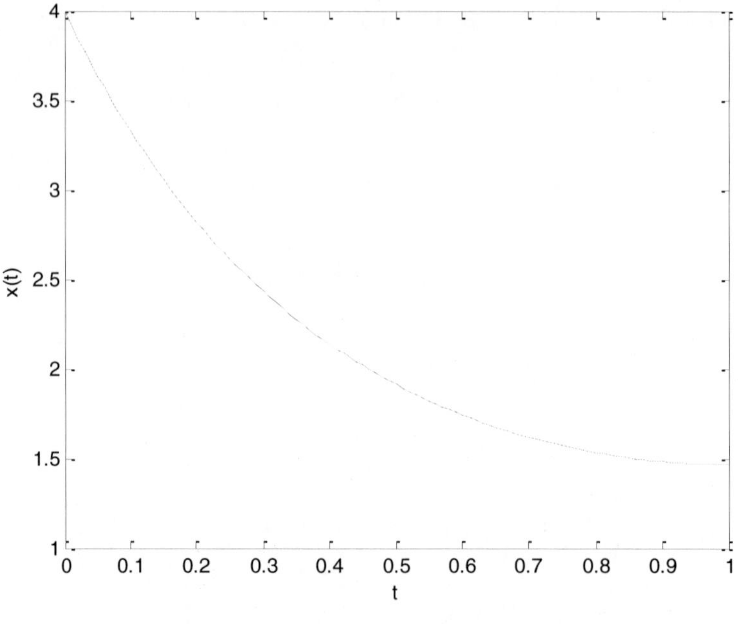

Figure 7.7. Optimum state vectors of Example 7.4 for $n = 102$.

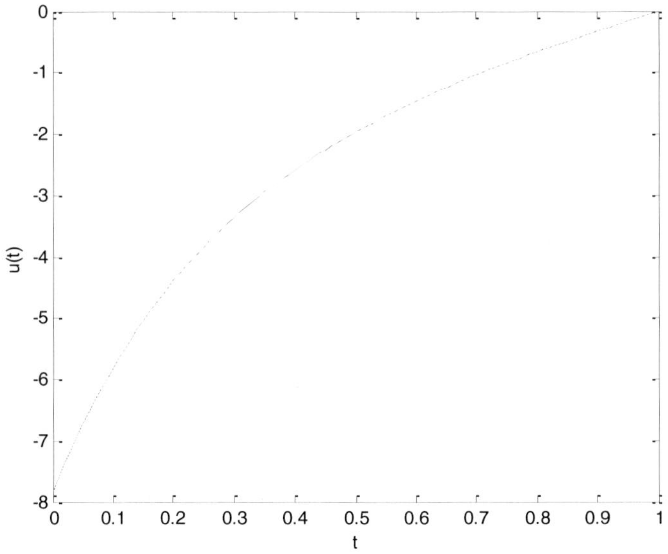

Figure 7.8. Optimum control vectors of Example 7.4 for $n = 102$.

$$\boldsymbol{x} = (4, 3.922888015459106, 3.848168398143530, \dots, 1.467530434143723)_{1\times 101}$$

with two iterations of the CAM (see Figures 7.7 and 7.8).

The results of the Examples 7.1, 7.2, 7.3 and 7.4 demonstrate that the output of the DISOPE algorithm was global based on Proposition 7.2.

Here, we investigated the application of the CAM for functionals successfully. The output of CAM by using UVCT was global based on the globality of CAM. Furthermore, the CAM has this property when it is used with a local search as a combination, converges to a global solution. This chapter demonstrated that the combination of CAM and DISOPE algorithm converges to a global solution based on Proposition 7.2. Experimentally, Examples 7.1, 7.2, 7.3 and 7.4 show the validation of this globality. Particularly, If we use CAM as a global search together with a local search, and if the output of the local search was global in the first place, then the output of the combination would also be the same value and hence global.

7.4. Some Suitable Suggestions for Improving the Combination Algorithm's Numerical Results

For further research in the application of CAM on functionals for decreasing the time taken for convergence the following suggestions are presented,

(i) For decreasing the number of vectors combinations in UVCT, it is more appropriate to substitute the vectors involved in the performance index. Particularly, in some examples, it is better to use only the state vectors (or control vectors) combinations to converge to the global optimum with reference to the constraints and the objective functionals.

(ii) Depending on the problems' forms it is better to find the suitable number of partitions experimentally in discretized OCP which is a major factor for converging to global solution rapidly.

(iii) In some examples, selecting the suitable stopping criterion shall help CAM to converge to the best solution in OCP.

References

[1] Bagirov, A. & Rubinov, A. (2000). "Global minimization of increasing positively homogeneous functions over the unit simplex", *Annals of Operations Research*, *98*(1-4), 171-187.

[2] Bagirov, A. M. & Rubinov, A. M. (2003). "Cutting angle method and a local search", *Journal of Global Optimization*, *27*(2-3), 193-213.

[3] Roberts, P. (1999). "Stability properties of an iterative optimal control algorithm", in *Proceedings of 14th IFAC World Congress on Automatic Control,* 269-274.

[4] Bryson, Jr. A. E. (1996) "Optimal control-1950 to 1985", *Control Systems, IEEE*, 16(3), 26-33.

AUTHORS' CONTACT INFORMATION

Dr. Seyedalireza Seyedi
Post-doctorate
Dipartimento di Matematica, Università di Bologna,
40126, Bologna, Italy

Dr. Iraj Sadegh Amiri
Assistant Professor
Computational Optics Research Group,
Ton Duc Thang University, Ho Chi Minh City, Vietnam
Faculty of Applied Sciences, Ton Duc Thang University,
Ho Chi Minh City, Vietnam
Email: irajsadeghamiri@tdt.edu.vn

Dr. Sara Chaghervand
Researcher
Department of Electrical Engineering,
Science and Research, Islamic Azad University,
Hamedan, Iran

Dr. Volker J Sorger
Department of Electrical and Computer Engineering,
The George Washington University,
Washington, D.C. 20052, USA

INDEX

A

abstract convex (AC), v, vii, 1, 2, 3, 6, 8, 9, 10, 12, 13, 15, 16, 17, 20, 23, 27, 34, 44, 51, 52, 53, 58, 59, 61, 63, 64, 65, 66, 68, 73, 74, 75, 76
algorithm, vii, 1, 5, 7, 8, 9, 12, 13, 15, 19, 21, 22, 23, 25, 27, 28, 29, 30, 31, 34, 77, 81, 83, 84, 85, 86, 91, 94, 98, 99, 103, 104, 123, 130, 131, 133, 134, 135, 137, 138, 139, 140, 142, 144, 146, 148, 149, 151
arithmetic, 5

B

Banach spaces, 50

C

calculus, 1, 9, 16, 19, 78
complex numbers, 48
complexity, 27
computer, 5, 17
computer simulations, 17
conjugate gradient method, 2, 19
construction, 3, 27, 39, 44
convergence, viii, 5, 21, 22, 29, 95, 100, 150
convex-along-rays (CAR), vii, 2, 3, 4, 6, 9, 12, 13, 16, 20, 27, 34, 51, 52, 58, 61, 69, 70, 71, 74
cutting angle method (CAM), v, vii, 1, 2, 4, 6, 7, 8, 9, 10, 12, 15, 16, 17, 20, 21, 23, 27, 29, 30, 34, 82, 93, 94, 95, 96, 97, 98, 100, 101, 105, 107, 115, 117, 120, 123, 129, 130, 133, 134, 135, 137, 138, 139, 140, 142, 144, 146, 147, 148, 149, 150

D

derivatives, 2, 134
discretization, 9, 28, 95, 96
dynamic integrated system optimization and parameter estimation (DISOPE), viii, 8, 9, 12, 15, 16, 27, 28, 29, 31, 34, 123, 130, 134, 135, 137, 138, 139, 140, 144, 146, 148, 149
dynamical systems, 12, 20, 30

E

engineering, 5, 30
Euclidean space, 12, 33

F

fermentation, 31
function values, 104
functional analysis, 1, 3, 8, 9, 46, 58

G

generalizability, vii, 33
geometry, 51
gradient descent method (GDM), 29, 134

I

increasing positively homogeneous (IPH), vii, 3, 4, 6, 9, 12, 13, 16, 17, 20, 21, 27, 30, 34, 52, 55, 56, 70, 99, 150
induction, 33, 36, 77
inequality, 46, 47, 67, 87
inheritance, vii, 33, 34, 43, 52, 86
integrated system optimization and parameter estimation (ISOPE), viii, 28, 29, 31
iteration, 5, 23, 27, 85, 104, 111, 114, 137

L

laws, 28, 43, 79, 82
linear function, 3, 37, 38, 39, 48, 49, 53, 54, 56, 58, 59, 61, 62, 63, 64, 67, 68
linear programming, 83
linear quadratic regulator (LQR), vii, 7, 8, 29, 94, 134, 139

M

mapping, 49
mathematics, 6, 33, 50, 78
matrix, 24, 50
models, vii, 7, 8, 17, 31, 94, 134

N

nonlinear systems, 13, 31, 131

O

optimal control problems (OCP), v, vi, vii, 1, 2, 6, 7, 8, 9, 10, 12, 27, 28, 29, 34, 77, 83, 91, 93, 94, 95, 96, 98, 100, 105, 106, 107, 117, 119, 120, 122, 123, 129, 130, 133, 134, 135, 140, 143, 144, 145, 146, 147, 148, 150
optimization, vii, 1, 2, 3, 4, 5, 6, 7, 9, 10, 12, 13, 15, 17, 18, 19, 20, 27, 28, 29, 30, 31, 34, 50, 76, 77, 78, 79, 80, 81, 82, 83, 85, 86, 93, 94, 95, 99, 133, 134, 135, 137
optimization method, 12, 30, 78

P

parallel, 97
parameter estimates, 29
parameter estimation, 15, 31, 133
programming, 3, 4
proposition, vii, 36, 57, 68, 72, 74, 75, 138

R

real numbers, 34, 47, 76
reality, 13, 17, 31, 131, 134
recombination, 86
researchers, 9, 28
roots, 6, 84

S

science, 17, 31, 78
scope, 9
Seyedi-Rohanin model (SRM), 37, 38, 39, 40, 41, 42, 43, 44, 45, 52, 69, 82
solution, vii, 4, 5, 6, 7, 8, 9, 17, 28, 29, 34, 83, 85, 86, 87, 91, 93, 94, 95, 98, 107, 119, 123, 129, 130, 134, 135, 137, 138, 139, 142, 144, 148, 149, 150
state(s), 16, 28, 29, 58, 63, 78, 96, 97, 99, 107, 116, 118, 121, 124, 125, 126, 127, 128, 130, 134, 142, 145, 146, 148, 150
strategy use, 5
structure, 15, 134

T

techniques, 1, 19, 29, 86, 95
topology, 34
trajectory, 89, 90
translation, vii, 7, 62, 94

U

unit vectors combinations technique (UVCT), viii, 9, 12, 95, 96, 97, 120, 130, 139, 142, 144, 146, 148, 149, 150

V

validation, 10, 138, 149
variables, 4, 18, 87
variational problems (VPs), 1
variations, 1, 9, 16, 19, 78
vector, 3, 20, 21, 23, 24, 33, 34, 46, 47, 48, 49, 50, 53, 54, 56, 57, 58, 59, 62, 66, 67, 68, 72, 74, 75, 84, 93, 97, 98, 110, 112, 113, 114, 116, 118, 121, 124, 125, 126, 127, 128, 130, 131, 135, 136, 137, 138, 143